The Anticipatory Design Playbook

The Anticipatory Design Playbook is a practical and thought-provoking guide to crafting anticipatory experiences in an age defined by artificial intelligence (AI). Grounded in years of PhD research and real-world design leadership, Joana Cerejo offers a comprehensive framework for designing AI-driven systems that align with user intent, respond to evolving behaviors, and remain transparent, adaptive, and human-centered.

Unlike most resources focused solely on the technical side of AI, this book bridges design, behavioral science, and machine learning to equip designers, product leaders, and technologists with the tools to shape intelligent systems (before those systems negatively shape us). From foundational concepts in AI, machine learning, and data science to behavioral theory, ethical design, and foresight methods, Cerejo introduces a new design language for systems that think ahead.

Readers will learn to identify key AI patterns, interpret mental models, and strike a balance between autonomy and automation across predictive, conversational, and agentic systems. The book introduces a structured process—anticipate, imagine, shape—that translates complex concepts into actionable UX strategies. Whether you're designing AI assistants, hyperpersonalized services, or adaptive platforms, this playbook offers practical methods for designing systems that anticipate needs, reduce friction, and support long-term behavior change.

More than a framework, *The Anticipatory Design Playbook* urges us to reclaim our role in shaping the future of intelligent systems. It is a call to leave behind the illusion of certainty that forecasting provides. It invites us to design with foresight—to create systems that are not just efficient, but context-aware, participatory, and resilient to change.

Joana Cerejo, PhD in Digital Media from the University of Porto, with over a decade of experience, is a UX lead and researcher specializing in human-centered AI and anticipatory systems. She was nominated for the 2021 Women in AI Awards by VentureBeat. With a background spanning UX research and strategy, behavioral science, and data-driven design, she helps organizations craft intelligent experiences that prioritize transparency, adaptability, and user autonomy. Since 2012, she has taught at universities and technical schools, mentoring new designers and sharing her knowledge.

The Anticipatory Design Playbook
A UX Guide to Design AI-Driven Experiences

Joana Cerejo

CRC Press
Taylor & Francis Group
Boca Raton London New York

CRC Press is an imprint of the
Taylor & Francis Group, an **informa** business

Designed cover image: Joana Cerejo

First edition published 2026
by CRC Press
2385 NW Executive Center Drive, Suite 320, Boca Raton FL 33431

and by CRC Press
4 Park Square, Milton Park, Abingdon, Oxon, OX14 4RN

CRC Press is an imprint of Taylor & Francis Group, LLC

© 2026 Joana Cerejo

ISBN: 978-1-041-07913-2 (hbk)
ISBN: 978-1-041-07910-1 (pbk)
ISBN: 978-1-003-64280-0 (ebk)

DOI: 10.1201/9781003642800

Typeset in Times
by Apex CoVantage, LLC

Contents

Acknowledgments

This book is the result of many years of research, reflection (and yes, occasional despair), and retrospective questioning of all my life choices. It would not exist without the support, inspiration, and encouragement of so many people who kept me going—sometimes with coffee, sometimes with wisdom, and often with both.

I am deeply grateful to my PhD supervisor, Professor Miguel Carvalhais, whose guidance and patience challenged me to think more critically, write more clearly, and occasionally accept that a good idea sometimes needs 20 more footnotes.

To my academic peers, thank you for the stimulating conversations, thoughtful questions, and shared confusion that helped shape the foundations of this work.

To the brilliant minds who shaped how I understand today's design challenges, Aaron Shapiro, Barry Schwartz, Roberto Poli, and B.J. Fogg, thank you for offering the kind of thinking that makes you highlight entire pages and rethink your project from scratch (in the best possible way).

To my colleagues in the design and AI communities: your real-world challenges, insights, and provocations reminded me why this work matters. And to those who gave early feedback, shared stories, or graciously stress-tested my ideas before they had a proper name, thank you. I'm especially grateful to the friends and fellow peers who kept cheering from the sidelines: Angela Ribeiro, Catarina Garcia, Margarida Carvalho, Marta Pinto, and Paulina Fonseca.

To my friends, thank you for your unwavering belief in me, even during the long, quiet months (okay, years). Your patience, humor, and grounding presence made the impossible feel just a little less impossible. And to my purr managers, Hannah and Kiko, thank you for your constant input, vigilant oversight, and for reminding me, every single day, that no deadline is more important than purring time.

And to the most important person in the world, my lighthouse, thank you for your unconditional support, endless encouragement, and for enduring this long academic and publishing voyage, often at the cost of our evenings, weekends, and precious time together. This book belongs to you as much as to me. I dedicate it to you, Nuno Domingues, with love and deep gratitude.

Finally, to everyone who believes that intelligent systems must be designed with empathy, foresight, and care, this book is for you. Thank you for choosing to spend time with my ideas.

Preface

In 2016, I found myself immersed in a machine learning (ML) project for the first time. What should have been an exciting challenge quickly turned into an overwhelming struggle. A labyrinth of unfamiliar terms, complex algorithms, and technical concepts that felt beyond my reach. As a designer, I searched for accessible guidance, only to realize that there were no resources tailored to our perspective and needs. The human-centered AI frameworks we see today simply didn't exist back then.

Determined to bridge this gap, I knew I needed a deeper, more structured understanding to truly support my users. That realization led me to pursue a more formal education, one that would equip me with the knowledge to navigate and influence AI-driven experiences.

Fast-forward to today. This book is the culmination of my PhD research—years of studying, observing, and experiencing firsthand the rise and fall of many AI-driven services. It introduces the first comprehensive approach to designing anticipatory experiences that align with user intent and mental models, making systems more intuitive, effective, and user-centered.

Whether you're a designer or not, this book is an invitation to step into the design room of tomorrow, where products predict, services adapt, and experiences unfold before the user even acts. Anticipatory systems represent not just a technological shift but a philosophical one. They challenge us to design for the "next," not just the now. That shift requires responsibility.

Jakob Nielsen once warned that while UX professionals focus elsewhere, engineers are shaping the future of AI-driven experiences. The result? Systems that are technically functional but experientially broken. The consequence? Designers are missing the opportunity to shape AI in ways that create meaningful societal and economic impact.

It's time to change that! This book is my response to that call.

I've structured it to serve a dual purpose: as a framework and as a field guide. You'll find theory backed by practice, heuristics rooted in cognitive science, and frameworks aligned with design realities. By grasping the fundamental principles of AI and ML at a high level, designers can work more effectively with engineers and data scientists. This understanding helps set realistic expectations, enhances communication, and ultimately leads to improved design outcomes.

This book is an invitation to reclaim our role in shaping AI and ensure that its future is not just intelligent but deeply human-centered.

THIS BOOK IS NOT JUST FOR DESIGNERS

While this book is grounded in design, it's written for anyone shaping or evaluating AI-driven experiences. That includes product managers, engineers, strategists, behavioral scientists, and policymakers. The complexity of anticipatory systems demands cross-functional literacy—technical fluency alone isn't enough, nor is creative intuition.

To support this, I include an onboarding chapter for readers less familiar with AI. It demystifies key terms—ML, data science, and neural networks—so you don't need a technical background to participate in the conversation. If you can frame a user problem, imagine a better path, or challenge an opaque algorithm, this book is for you.

IN A NUTSHELL

This book is not a blueprint—it's a navigational tool for an emerging discipline: **anticipatory design**. It offers a language, a structure, and a set of tools to help you design AI-driven experiences that are not just predictive, but genuinely anticipatory—systems that evolve with their users, respond with care, and align with long-term human intent.

You'll find a blend of theory and application. Early chapters establish key concepts from artificial intelligence (AI), ML, behavioral science, and systems thinking, without requiring a technical background. Middle chapters explore the psychological, temporal, and ethical dimensions of anticipatory systems—how users form mental models, how we align with intent, and how methods like forecasting and backcasting help us design for what lies ahead. These perspectives shift the focus from what systems *can* do to what they *should* do—and when.

Later chapters introduce a structured framework—*anticipate, imagine, shape*—organized into cognitive UX chunks, with actionable heuristics, research prompts, and design methods for creating adaptive, human-centered systems.

The book is modular by design. You can read it linearly to follow the whole arc from theory to implementation, or dive into specific chapters to address immediate challenges. UX designers will find patterns and processes, product leaders will gain strategic insight, and researchers and technologists will gain clarity on the human factors that predictive systems often overlook.

Above all, this book is an invitation to design with the future in mind—not just to optimize the next click but to shape long-term behaviors, relationships, and decisions. As anticipatory systems become more powerful, our design questions must become more principled: are we aligning with user intent, or just predicting behavior? When systems act on our behalf, who defines what "helpful" looks like? What's being removed in the name of convenience? How do we preserve agency in systems that think ahead? Whether you're here to build, question, or reimagine, you're in the right place.

PART I

Setting the Stage

Understanding Artificial Intelligence, Data Science, Machine Learning, and Deep Learning: The Layers for Anticipatory Systems

Foundational Concepts of Artificial Intelligence and Machine Learning

1

This chapter introduces key concepts in artificial intelligence, data science, and machine learning, setting the stage for a deeper exploration of how these technologies reshape industries and design. Before delving into technical aspects, let's first reflect on the evolving relationship between humans and machines.

From the moment we are born, we are surrounded by vast amounts of data. Our senses serve as tools for gathering data and continuously capturing sounds, sights, smells, and textures. Our brain processes this raw data to form perception, knowledge, and communication. We don't just experience reality—we record it, share it, and transform it into meaning. This ability to collect, structure, and communicate information is essential to human progress and evolution.

Now, imagine a machine trying to do the same. While machines process vast amounts of data, their approach is fundamentally different. Instead of relying on intuition, experience, or common sense, machines depend on mathematical patterns. Through machine learning (ML) and its powerful subset, deep learning (DL), computers analyze data, detect patterns, and make predictions—all without being explicitly programmed for every scenario. Unlike humans, machines operate purely on statistics, lacking intuition and context.

DOI: 10.1201/9781003642800-2

1.1 RULE-BASED VERSUS LEARNING-BASED SYSTEMS

To understand how AI is transforming our world, let's consider two different approaches to problem-solving: one that follows rigid instructions and another that learns from experience.

Let's begin with the **rule-based system**. Think of it as a well-documented recipe for the perfect burger. Every step is explicitly outlined: toast the bread for precisely 30 seconds, season the burger with a precise amount of salt, and flip it at the two-minute mark. The result? It is consistent, predictable, and repeatable—just like traditional software that operates on predefined rules.

Now, consider a **learning-based system in contrast**. Instead of following a rigid recipe, this approach resembles a chef who has tasted thousands of burgers, experimented with flavors, and continuously refines their technique based on experience and feedback. Rather than following a fixed set of instructions, this system learns from data and adapts over time. In AI terms, this means that the system identifies patterns in the data through a process called **feature extraction** (such as ingredients, textures, and flavors) and builds a model that evolves as it processes more information.

In traditional programming, developers write explicit rules for the system to follow. In contrast, ML systems learn from data, identify patterns, generate insights, and adapt to new inputs. This adaptability is one of the primary reasons why ML is transforming industries, including design.

Arthur Samuel defined ML in the 1950s as "the field of study that gives computers the ability to learn without being explicitly programmed" [1]. Since then, advances in computing power, storage, and algorithms have made ML widely applicable across various industries.

Unlike traditional software, which follows fixed instructions, ML mimics continuous learning. It enables machines to adapt and refine their behavior based on new data, similar to human learning. This adaptability is why ML is now the most prominent form of AI. Recent breakthroughs in computational capabilities and algorithm development have created ideal conditions for ML to flourish, unlocking a range of applications that were once thought to be impossible.

1.1.1 Artificial Intelligence and the Future of Design

From personalized recommendations in e-commerce to AI-assisted design tools, designers no longer create static systems; instead, they create dynamic, ever-evolving experiences that respond to user behavior in real time. This shift requires technical literacy and a deep understanding of ethics, transparency, and user agency in AI-driven designs.

Today, AI is not just a tool; it is a **design material** that facilitates the creation of adaptive, dynamic experiences. By incorporating AI capabilities into user journeys, designers can develop systems that anticipate user needs, predict outcomes, and deliver personalized, intuitive experiences. To accomplish this, designers must grasp the principles of AI, ML, and DL and apply them ethically to ensure transparency, autonomy, and trust in the resulting systems. Just as designers once mastered physical materials like paper, ink, and metal and later adapted to digital interfaces, they must now embrace **adaptive, AI-driven experiences** as a vital part of their toolkit.

Some fear that AI will replace designers, but I believe that designers who embrace AI will be at the forefront of innovation. AI is not here to replace human creativity but to amplify it. Designers who integrate AI into their workflows will unlock new creative possibilities, optimize user experiences, drive innovation, and make a meaningful societal and economic impact.

1.2 WHAT IS ARTIFICIAL INTELLIGENCE?

Data lies at the core of AI, driving decision-making across industries and in our daily lives. At a higher level, this transformation is rooted in **computer science**—a broad field that encompasses everything from algorithms to systems design. Within this field, **data science** emerges as a discipline focused on extracting valuable insights from data

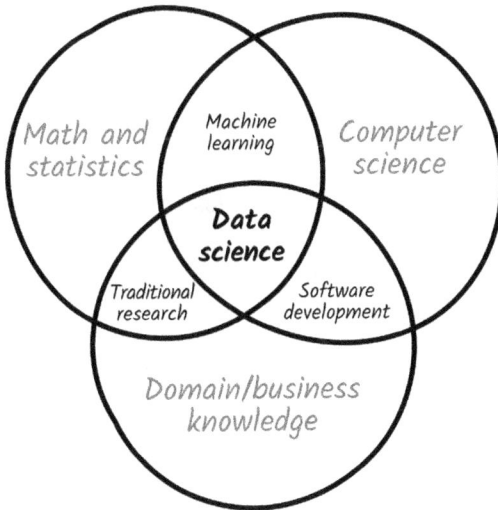

FIGURE 1.1   Venn diagram showing data science at the intersection of math and statistics, computer science, and domain or business knowledge.

Computer science   Data science   Artificial intelligence   Machine learning

FIGURE 1.2 Matryoshka Dolls analogy to illustrate the relationship between computer science, data science, artificial intelligence, and machine learning.

using computational, statistical, mathematical, and scientific methods. One powerful application area of data science is AI, which enables machines to perform tasks that traditionally require human intelligence, such as decision-making, problem-solving, and creativity. A key subset of AI is ML, which enables systems to learn from data and improve over time without requiring explicit programming.

AI is at the heart of this data-driven transformation, **enabling** machines to perform tasks that require human-like intelligence and cognitive abilities. Just as the **industrial revolution** introduced machines that replicated physical labor, AI is leading a new era where machines emulate **cognitive functions,** handling tasks such as decision-making, problem-solving, and even creativity.

AI is often synonymous with ML, but it's essential to understand that ML is only one part of the broader fields of data science and computer science. Recognizing how AI relates to these disciplines is crucial, as terms like ML and data mining often overlap and can be easily confused.

Although AI encompasses various subfields, ML stands out as the primary force driving most practical applications. Nevertheless, it is important to note that ML is a relatively small component of the broader domains of data science and computer science. Grasping how AI subfields relate to their parent disciplines and associated areas is crucial, as you will often encounter these terms in your work. Additionally, it can be challenging to distinguish between similar concepts, particularly ML and data mining.

1.2.1 Data Mining Versus Machine Learning

ML has a twin sister—**data mining**. Like any twins, they look very similar but have entirely different personalities. Both aim to uncover patterns in large datasets, but their goals diverge significantly. Data mining focuses on discovering insights and

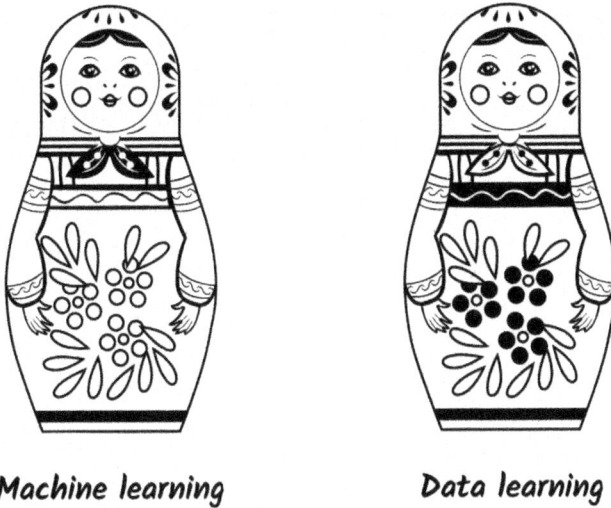

Machine learning   **Data learning**

FIGURE 1.3   Two similar Matryoshka dolls: machine learning versus data mining.

relationships within the data without attempting to predict future events. For instance, analyzing supermarket purchase data might reveal that bread and butter are often bought together. However, data mining does not aim to forecast future purchases; it simply uncovers patterns in the present.

In contrast, ML focuses on predictions. It utilizes the same data to anticipate future behaviors, such as when a customer is likely to buy bread again based on their prior purchases. While data mining and ML employ similar tools and methods, their emphasis differs. Data mining reveals what is happening now, whereas ML forecasts what is likely to occur next. To better understand the distinction between them, let's take a closer look at each of their focuses.

Think of data mining as a detective investigating a crime. The detective's goal is to piece together the details of what happened—gathering clues, examining evidence, and reconstructing events to understand the past. The detective is focused on uncovering the *why* behind the crime, but there's no need to predict future events. Data mining's primary objective is to explore the data and its relationships, rather than to forecast what might come next.

Now, imagine ML as a different detective working with the same data. This detective's mission is to predict future crimes by taking a prediction-driven approach. Consider the 2002 film *Minority Report*, where PreCrime detectives predict crimes before they occur. Similarly, ML analyzes historical crime data to predict where and when future crimes are likely to occur. By identifying patterns in the data, the ML detective makes predictions and takes action to prevent future incidents. The goal here is not to understand the past but to predict and intervene in upcoming occurrences.

1.2.2 Machine Learning Versus Deep Learning

DL is a more advanced subset of ML. The primary distinction between them lies in their **level of autonomy**. While ML still requires human intervention for tasks like feature extraction, DL takes it a step further by enabling machines to train themselves with minimal human oversight.

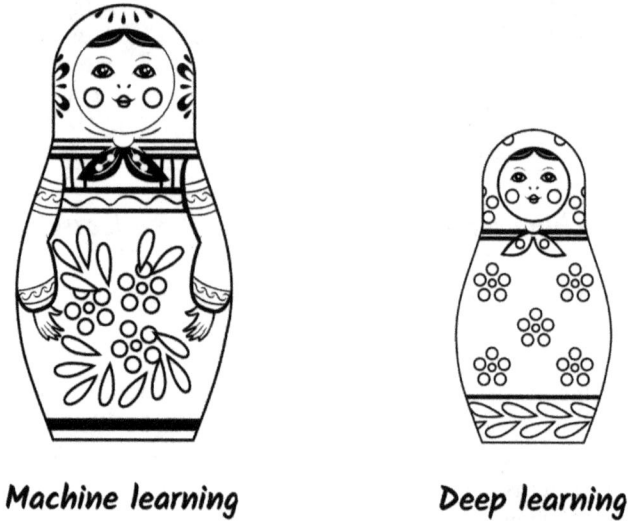

Machine learning   **Deep learning**

FIGURE 1.4   A Matryoshka doll inside machine learning-deep learning doll.

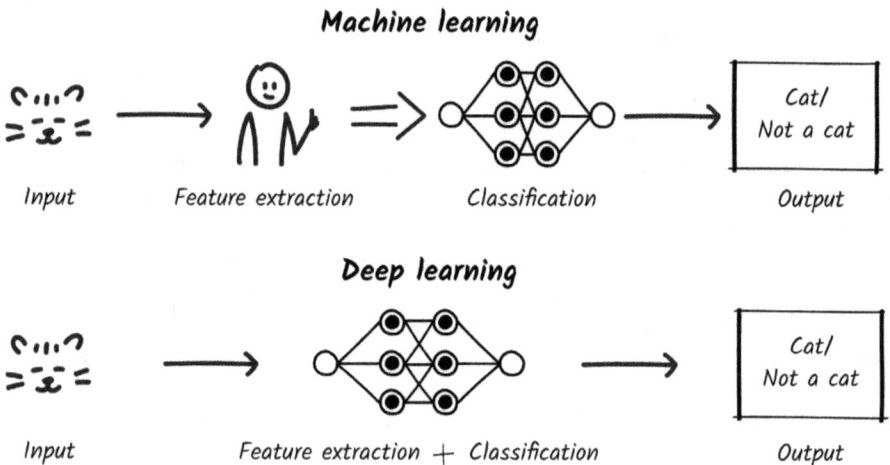

FIGURE 1.5   Comparison of machine learning and deep learning workflows.

Remember the burger example? One step, feature extraction, involved teaching the machine which elements we liked most in each burger. The main difference between ML and DL is that, in ML, a human performs this task, whereas in DL, the algorithm learns autonomously to extract these features by itself.

Therefore, the critical difference between ML and DL is their level of autonomy:

- **ML** requires a human to help the computer identify important features.
- **DL**, on the other hand, is more "independent." It learns to recognize these features independently, without human intervention.

1.2.2.1 Connectionism

All the previous definitions lead to connectionism, an approach to studying human cognition through mathematical models known as connectionist networks or artificial neural networks (ANNs). These networks are often modeled after highly interconnected neuron-like processing units.

A neural network is a digital reconstruction of connections in the brain. DL relies on neural networks to learn the underlying features of a dataset. An ANN mimics the way neurons in the brain process information. Artificial neurons, also known as nodes, are organized into layers and operate in parallel. When a node receives a numerical input, it processes the signal and passes it on to other connected nodes. Like the human brain, neural networks enhance their pattern recognition and learning capabilities through reinforcement.

Machine learning   **Deep learning**   **Neural networks**

FIGURE 1.6 Visual metaphor using Matryoshka dolls to show that neural networks are part of deep learning, which is itself part of machine learning.

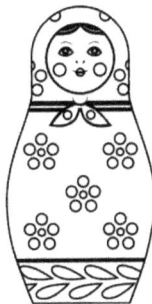

Neural networks can identify much more complex and abstract patterns than traditional ML methods. This is why we can place another Matryoshka doll within the ML and DL category: **neural networks**. Neural networks enable machines to learn tasks directly from raw data, eliminating the need for manual feature engineering.

1.3 THE ROLE OF DEEP LEARNING IN AI

Neurons, often referred to as brain cells, are the building blocks of our nervous system, allowing us to sense, think, and act. In the nineteenth century, Spanish physician Santiago Cajal was the first to identify neurons by meticulously staining and examining thin slices of brain tissue [2]. Only a century later, by the 1950s, propelled by an expanding understanding of the brain, scientists began experimenting with artificial neurons, developing ANNs that loosely mirrored their biological counterparts.

The 2018 Turing Award, often regarded as the "Nobel Prize of computing," was presented to three researchers who established the foundation for today's AI advancements. Yoshua Bengio, Geoffrey Hinton, and Yann LeCun—commonly known as the "godfathers of AI"—emphasized that DL emerged as a solution to the shortcomings of traditional ML techniques, which struggled to manage raw, unprocessed data. Developing a pattern recognition system or ML model for decades has required specialized expertise and intensive engineering, rendering the process time-consuming and complex. DL filled this void by automating the feature extraction process, enabling machines to learn directly from raw data. Before this breakthrough, creating effective feature extractors (such as the burger example) was a meticulous craft that required deep domain expertise and considerable trial and error. DL democratized ML and empowered models to tackle more complex problems by removing this manual step.

In sum, DL refers to using multiple layers in neural networks to process large amounts of complex, unstructured data. Each layer progressively extracts and refines information, enabling the model to learn directly from the data without requiring human intervention to design specific feature extractors.

1.3.1 Black Box

The term "black box" originates from DL. The numerous layers and automatic decisions made during this process make it incredibly difficult, if not impossible, to understand how the machine makes a particular decision. We are aware of what data is entered and what output is generated, but the internal decision-making process remains unclear. To better understand this, let's take a closer look at what makes up a black box in DL. In a neural network, there are three primary components:

- **Input layer**: This is where the data is entered—in this case, images of two types of mammals, hedgehogs and squirrels.

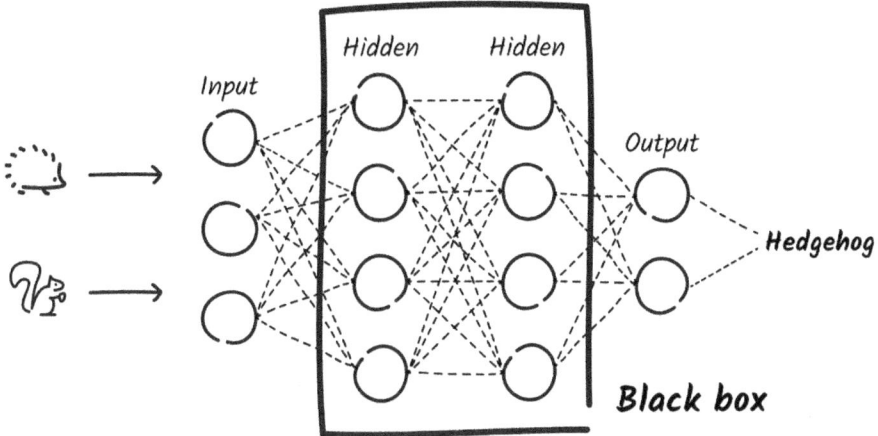

FIGURE 1.7   Illustration of what constitutes the black box.

- **Hidden layers** are the "black box" of DL. In these layers, the data undergo multiple transformations, some of which humans cannot easily interpret.
- **Output layer**: This provides the final classification, for example, determining whether the input is "a hedgehog" or "a squirrel."

As data flows through the hidden layers, the model progressively learns by processing the data layer by layer, as though viewing it through multiple lenses. In the first layer, the model identifies basic shapes such as "Does it have visible ears?" or "Does it have a tail?" The second layer begins analyzing more detailed features, such as "What is the animal's color?" or "Is the shape round or narrow?" This process continues, with each layer refining the understanding, allowing the model to make accurate predictions, such as correctly identifying a hedgehog. Each layer deepens the analysis, transforming simple data into more complex features, which helps the model learn abstract representations and make accurate decisions.

Despite the complexity of DL, there are two primary specializations:

- Convolutional neural networks (CNNs)
- Recurrent neural networks (RNNs)

CNNs specialize in processing image and visual data. These networks excel at identifying spatial patterns such as shapes and textures, making them ideal for tasks like object recognition, image classification, and computer vision. In medical applications, CNNs are invaluable tools, such as in SkinCheck, where they analyze images of moles and skin spots. By recognizing visual patterns characteristic of skin cancers, CNNs help detect the disease more accurately and earlier, significantly improving diagnostic and treatment outcomes.

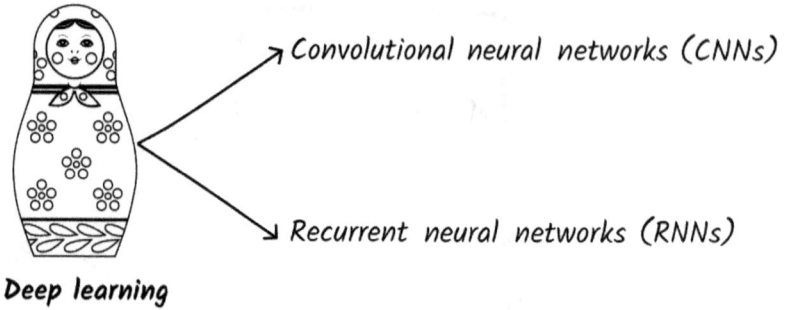

FIGURE 1.8 Specializations in deep learning: CNNs and RNNs.

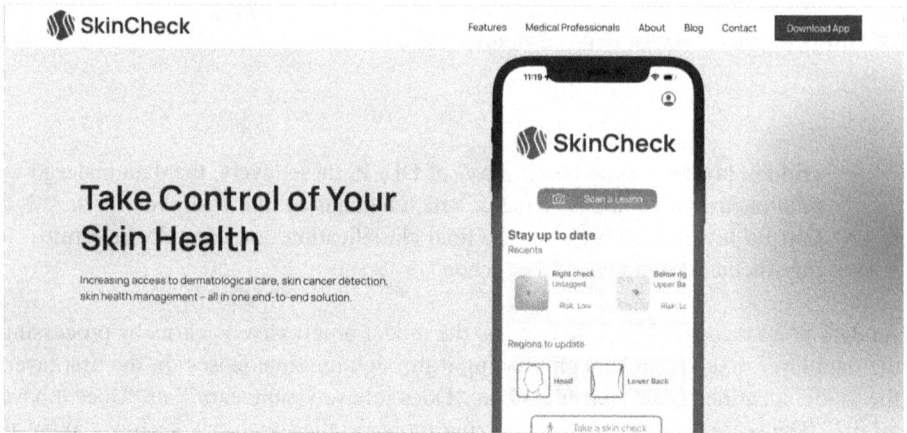

FIGURE 1.9 A 2025 screenshot of the SkinCheck website at www.skincheck.health.

Moreover, CNNs are crucial in the fight against serious illnesses, such as cancer. These neural networks analyze medical images, such as X-rays, CT scans, and MRIs, to identify early signs of tumors, thereby increasing the chances of early diagnosis and effective treatment.

RNNs are designed to handle sequential or temporal data, including text, audio, and time series information. Unlike CNNs, which process data independently in each layer, RNNs have internal loops that allow them to retain information from previous steps. This "memory" capability enables RNNs to maintain context, which is crucial for tasks like machine translation, text generation, and time series forecasting.

ChatGPT serves as an excellent example of an RNN application that utilizes internal memory to understand and respond coherently to dialogue. By analyzing each word in a sentence, the system can predict the next word or interpret the overall meaning based on the context, thereby providing a more natural flow of conversation.

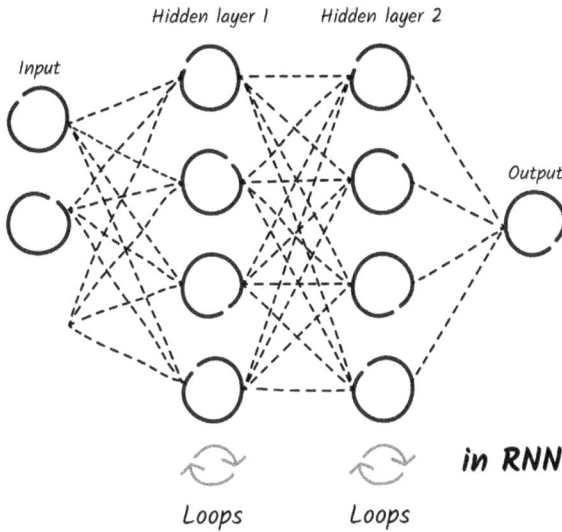

Hidden layer 1   *Hidden layer 2*

Input

Output

in RNN

Loops   *Loops*

FIGURE 1.10   Illustration of RNN loops.

In conclusion, when people refer to AI or companies that market AI-driven products, they often discuss ML; when they mention ML, they frequently mean DL. What I hope you take away from this chapter is that ML, at its core, is fundamentally pattern recognition—a subset of AI. DL builds on this, serving as an enhanced version of pattern recognition, enabling systems to address more complex and nuanced challenges.

1.4 ONLINE VERSUS OFFLINE LEARNING

As we have seen, ML models are typically trained on a dataset before being deployed in real-world applications. However, the way these models are released into the world can vary, primarily between online and offline learning methods. The key difference lies in how data is processed and updated over time.

In online learning, the model continuously updates itself as new data arrive, adapting in real time and adjusting its parameters accordingly. Consider it like a designer constantly refining their work based on immediate user feedback. A prime example of online learning is Waze.

Waze needs to adjust its routes based on real-time traffic data. As users drive and share data about traffic speed, jams, or accidents, the system immediately learns and recalculates the optimal routes. This constant update process makes navigation highly efficient. If Waze used offline learning, its maps and route suggestions would only update periodically, meaning it couldn't respond to sudden real-time changes, like a traffic jam or accident.

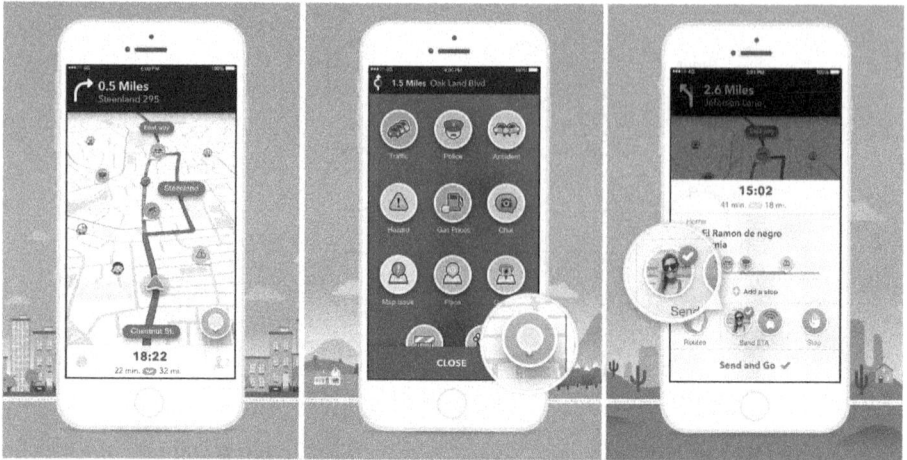

FIGURE 1.11   Waze mobile application illustrating the service's key features.

Offline learning, also known as batch learning, works by training the model on a large dataset all at once. Once the model is trained, it remains unchanged until it is retrained with new data. This approach is similar to a designer periodically creating and updating a fixed template. For example, *Spotify's Discover Weekly* feature analyzes user data every week to generate new, personalized playlists. The model is updated in batches, so it doesn't react instantly when a user listens to a new song; the change happens during the next update.

The offline learning model is known for its stability. It operates with a static dataset, making it ideal for applications that require consistent and reliable results. It is often simpler to implement than online learning, making it accessible in various scenarios. As a result, offline learning excels at identifying historical patterns and trends, offering a powerful tool for strategic decision-making.

1.4.1 Which Is Better?

Each method has its strengths and weaknesses.

Online learning: The model must make decisions quickly, as data streams in real time. This means that it might not always analyze all available data before updating. It adapts as new information arrives, but doesn't have the luxury of processing everything beforehand.

Offline learning: The model can analyze large datasets and make well-informed predictions. However, once trained, it doesn't adapt to new data immediately. Instead, it requires retraining to incorporate fresh information.

The choice between online and offline learning depends on several factors, like:

- **Nature of the data**: Whether the data is static or constantly changing.
- **Response time requirements**: Whether predictions need to be made in real time.
- **Available computing resources**: Online models may be more expensive to train and maintain due to their requirement for continuous updates.
- **Application objective**: Whether the goal is to identify historical patterns or make real-time predictions.

1.5 DATA LAKES VERSUS DATA WAREHOUSES

The choice of data repository depends on the ML model required for our service. Different models require different data storage solutions. The two primary types of data storage are **data lakes** and **data warehouses**.

Data lakes are designed to store raw, unprocessed data from various sources and formats. These repositories can support both online and offline learning models while accommodating a variety of data types and applications. In contrast, **data warehouses** are structured repositories optimized for analysis and reporting. They typically contain organized data and are often utilized for offline model training due to their data consistency.

Engineers interact with data in these repositories in a significantly different manner. This interaction, known as system transactions, refers to the specific events that an organization aims to track. For example, in a banking system, a transaction might involve transferring money between accounts, while in retail, it could include processing payments. Transactions represent small, discrete units of work within a system.

1.5.1 Online Transaction Processing

Data lakes often use data generated by **online transaction processing (OLTP)** systems, which manage real-time transactions such as those in online shopping. This data is transformed and cleaned before being used as input for online ML models. OLTP acts like a "supermarket checkout," quickly processing transactions and ensuring data accuracy. It tracks each sale or customer interaction while handling a high volume of simultaneous transactions.

1.5.2 Online Analytical Processing

Data warehouses, however, often leverage **online analytical processing (OLAP)** systems. These systems are designed for the complex, multidimensional analysis of large

FIGURE 1.12   Spotify marketing strategy leveraging OLAP data analysis [3].

volumes of historical data. OLAP is ideal for answering strategic questions, such as analyzing consumer behavior and trends.

A practical example is **Spotify**, which uses OLAP systems to analyze user content consumption patterns. This data informs features such as *Discover Weekly*, which personalizes the user experience and influences business strategies. Spotify also leverages OLAP to create bold and creative marketing campaigns based on the insights gathered from this type of analysis.

Unlike OLTP, which focuses on individual transactions (such as tracking every song played), OLAP organizes aggregated data along multiple dimensions to uncover long-term trends and patterns. Data warehouses provide quick access to optimized data for generating detailed reports and creating interactive dashboards.

1.5.3 How Do AI and ML Affect Designers?

AI and ML are reshaping the world, yet the role of designers in this transformation is crucial. Until now, engineers and data scientists have largely driven technological developments. However, the true potential of these technologies depends not only on algorithms but also on how people interact with them, and that's where designers play a vital role.

Designers can significantly influence how ML and AI technologies are conceived, developed, and applied to products and services. Thanks to their skills in empathy, systems thinking, and user experience, designers can shape user interfaces and experiences.

However, because of a lack of technical training, many designers are excluded from or only involved in the later stages of AI and ML development.

Challenges Designers Face:

- **Understanding AI**: A limited understanding of how AI works hinders designers from spotting opportunities to integrate AI into products. When designers lack knowledge of ML and DL, simple application opportunities are often overlooked.
- **Lack of technical language**: Designers often lack the technical language to collaborate effectively with engineers and data scientists. They may also not fully grasp AI's capabilities and limitations, leading to unrealistic expectations or missed opportunities.
- **Impact on design**: Many AI-based solutions fail to reach their full potential, not because of weak technology but because the design doesn't integrate effectively.

1.5.4 The Designer's Role in This Revolution

With great power comes great responsibility. Designing for AI and ML is different from traditional design due to the highly adaptive nature of these systems. Unlike pre-programmed systems, their behavior is derived from data, meaning the outputs may vary based on what the system learns. Designers must recognize that AI-driven systems are dynamic, and the outputs can be unpredictable. Flexibility in interface and experience design is essential.

Instead of designing based on fixed rules, designers must:

- Understand the **origin**, **quality**, and **bias** of the data.
- Design interfaces that **clearly convey how data is used**, promoting transparency to build user trust.
- Work closely with engineers, data scientists, and other technical stakeholders.
- Designers must be able to **communicate technical possibilities** and **balance user needs** with **technical constraints**.
- ML produces **probabilistic** results, not absolute ones. This means designers must be able to clearly handle errors and uncertainties, ensuring transparency and user safety.
- Designers must create systems that **allow user intervention**, such as correcting predictions, and ensure that systems evolve as new data is introduced.
- Design is an ongoing process that requires **constant monitoring** and regular updates to enhance usability and performance.

In summary, designing for AI and ML requires a new approach—one that is flexible and collaborative. This approach enables designers to tailor these technologies to create more effective user experiences and outcomes.

Too Long; Didn't Read (TL;DR)

This chapter lays the groundwork for understanding how artificial intelligence (AI), machine learning (ML), and deep learning (DL) function and how they're transforming design.

Humans and machines learn differently—Humans rely on intuition, context, and perception to make informed decisions. Machines, by contrast, learn through data patterns and statistical models—especially via ML and DL.

Rule-based versus learning-based system—Traditional (rule-based) systems follow explicit instructions. Learning-based systems adapt to data, evolving over time to make more accurate predictions and informed decisions.

Machine learning versus deep learning—ML requires human involvement in feature selection. DL automates this process through neural networks, allowing systems to learn complex patterns directly from raw data.

Neural networks and connectionism—Inspired by the human brain, ANNs process data in layered architectures, deepening understanding with each layer. DL uses these to automate and refine feature recognition.

The "black box" problem—DL models are powerful but opaque, difficult to interpret or explain due to the complexity of their inner workings. This has significant implications for transparency and trust.

Specialized Networks:

- **Convolutional neural networks (CNNs)** excel at analyzing visual data (e.g., SkinCheck for early detection of skin cancer).
- **Recurrent neural networks (RNNs)** handle sequential data, such as language (e.g., powering ChatGPT's conversational flow).

Online versus offline learning—The key difference lies in how data is processed and updated over time:

- **Online learning** adapts to new data in real time (e.g., Waze), making it ideal for fast, adaptive systems like streaming or live recommendations. It continuously updates but may sacrifice deep analysis due to limited processing time.
- **Offline learning** is trained in batches and updated periodically (e.g., Spotify's Discover Weekly), offering deeper insights from large datasets. It suits stable environments where periodic retraining is acceptable, like scheduled model updates.

Data Infrastructure—Data lakes versus warehouses:

- *Data lakes* are flexible and support the storage of raw, real-time data for online learning.
- Data warehouses are structured and best for historical and strategic analysis via OLAP.

The designer's role in the age of AI—Designers must go beyond aesthetics, understanding data flows, algorithmic behavior, and ethical implications. They should:

- Collaborate with engineers and data scientists.
- Make AI systems transparent, flexible, and trustworthy.
- Design for uncertainty, user control, and continuous adaptation.

Key AI Capabilities and Real-World Use Cases

2

> *With these foundational concepts in place, let's explore their practical manifestations across various industries. This chapter reflects on the statement, "All Machine Learning is Artificial Intelligence, but not all Artificial Intelligence is Machine Learning." To grasp this concept, the chapter examines key AI patterns and analyzes their diverse applications through concrete examples.*

In the previous chapter, we saw how ML has enabled computers to understand what we say, what we see, and even how we imagine, revolutionizing human–computer interaction. However, ML is just one part of AI. To fully grasp AI's potential, we must examine broader patterns of AI that extend beyond ML. One of the most significant areas of AI is computer vision.

2.1 COMPUTER VISION

Computer vision is a branch of AI that enables machines to interpret and process visual data, mimicking human visual perception. This capability allows computers to recognize objects, analyze images, and make informed decisions based on visual inputs. In essence, computer vision equips computers with a digital *"eye,"* enabling them to perceive and interpret the world around them.

2.1.1 How Does Computer Vision Work?

Understanding how computers process images is similar to understanding how the human brain interprets visual information. Just as neuroscience explores the complexities

DOI: 10.1201/9781003642800-3

FIGURE 2.1   Computer vision example [4].

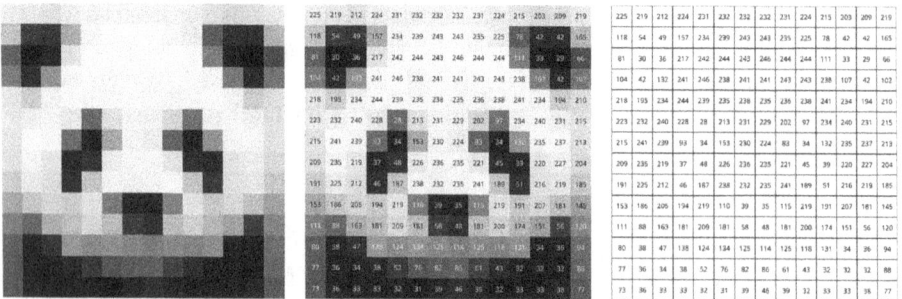

FIGURE 2.2   Image representation in pixels [5].

of human perception, ML models are trained to analyze and make sense of visual data. A computer interprets any image by following these steps:

- **Acquire an image** from a camera, video, or digital source.
- **Process the image** by analyzing its pixel grid.
- **Understand the image** by assigning numerical values (ranging from 0 to 255) representing pixel brightness.

2.1.1.1 Computer Vision Models and Capabilities

Computer vision encompasses various AI techniques that allow computers to detect, classify, and analyze images or videos. These capabilities have transformed industries ranging from healthcare to retail, as well as self-driving cars and security. The six most common computer vision patterns are as follows:

- **Image classification**: Assigns a label to an entire image (e.g., classifying an image as a "cat" or "dog").
- **Object detection**: Recognizes and pinpoints specific objects within an image.
- **Semantic segmentation**: Divides an image into meaningful regions by assigning a label to each pixel.
- **Image analysis**: Extracts insights from visual data, such as detecting anomalies in medical images.
- **Face detection, analysis, and recognition**: Identifies and verifies human faces.
- **Optical character recognition (OCR)**: Converts images of text into machine-readable text.

2.1.1.2 Examples of Computer Vision Applications

Computer vision aims to address key problems, including image classification, object detection, and semantic segmentation.

Image classification is a foundational technique in computer vision. An ML model is trained to categorize images based on their content. For example, in a traffic monitoring system, image classification can distinguish between various types of vehicles, including taxis, buses, and bicycles. A practical application of image classification is Pinterest, which uses this capability to automatically categorize and suggest related pins based on the content of the images users save. This technology powers its recommendation engine, significantly enhancing the user experience.

Object detection models classify individual objects within an image and pinpoint their locations using bounding boxes. For instance, a traffic monitoring system can utilize object detection to identify and track various types of vehicles. A well-known example is **Amazon Go** stores, which utilize object detection to track items that customers select from shelves, allowing for a checkout-free shopping experience. The system detects products in real time and automatically processes transactions.

Semantic segmentation is an advanced computer vision technique that classifies each pixel in an image according to the object it represents. For instance, a traffic monitoring solution might overlay a video feed with color-coded "masks" to distinguish different vehicles. Photoshop's "Select Subject" feature serves as a well-known example of semantic segmentation. This tool automatically isolates the main subject in an image, simplifying the process of editing or removing the background.

FIGURE 2.3   Object detection classifies objects within their location by using bounding boxes.

Computer vision problem types

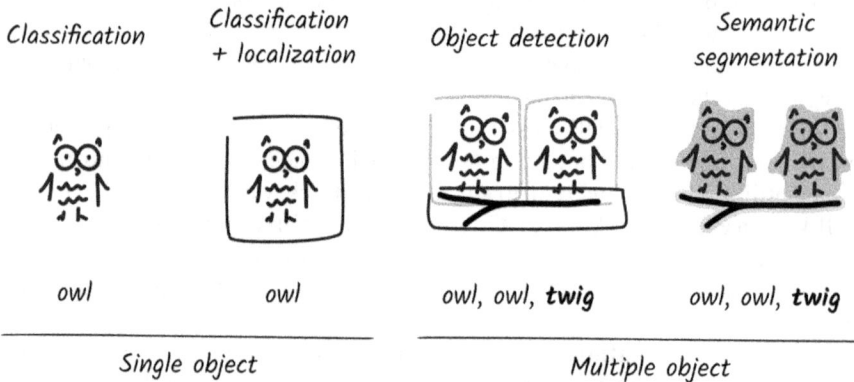

Classification	Classification + localization	Object detection	Semantic segmentation
owl	owl	owl, owl, **twig**	owl, owl, **twig**

Single object	Multiple object

FIGURE 2.4   Illustration of computer vision problem types.

Combining ML models with advanced image analysis techniques enables the extraction of information from images, the generation of descriptive captions, and automatic tagging of images for organization. A prime example is **Google Lens**, which allows users to analyze images through their smartphones or web browsers. Google Lens can identify objects, translate text, recognize landmarks, and even find similar products.

Face detection is a specialized type of object detection that identifies human faces within an image. When paired with facial geometry analysis, it can confirm identities. A popular example is **Apple's Face ID**, which utilizes face detection, analysis, and recognition to unlock iPhones and authenticate transactions securely.

Finally, we have **OCR**, a technique that detects and interprets text in images. It is especially beneficial for extracting information from scanned documents, photographs of signs, or printed text. Adobe Acrobat Pro is a prominent example of OCR in action, with built-in capabilities that allow users to scan printed documents and convert them into editable and searchable PDFs. OCR is also widely used in accounting and finance, where it accelerates the collection of data from receipts, invoices, and other financial documents, dramatically reducing manual entry and streamlining bookkeeping processes.

2.1.2 Challenges and Considerations for Computer Vision

We have explored various computer vision applications, many of which focus on specific functionalities. However, real-world solutions often integrate multiple computer vision techniques to develop more sophisticated applications.

One such example is **Seeing AI**, a free app developed by Microsoft that utilizes computer vision to assist individuals who are blind or have low vision in understanding their surroundings. It leverages the phone's camera to recognize and describe objects, text, people, and more. Although Seeing AI has been around for several years, it remains a standout example of how multiple computer vision techniques can be combined to create a powerful assistive tool.

Despite its vast potential, computer vision also presents challenges that designers must address. The accuracy of image recognition models relies heavily on the quality

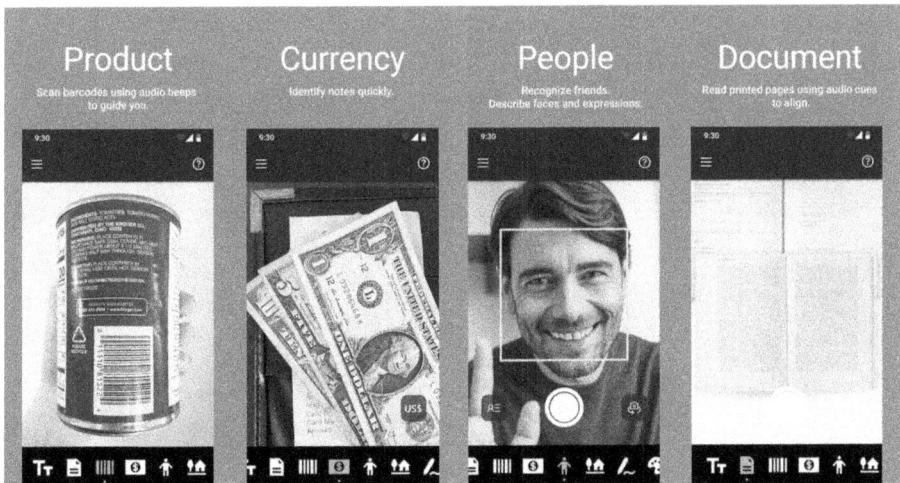

FIGURE 2.5   Microsoft Seeing AI app [5].

and diversity of the data on which they are trained. Biases in datasets can result in inaccurate or unfair outcomes, making it essential for designers to actively curate, test, and refine the data used in these applications. By prioritizing high-quality and diverse datasets while improving algorithms, we can ensure that computer vision serves users effectively and ethically.

2.2 NATURAL LANGUAGE PROCESSING

Natural language processing (NLP) is among the most transformative areas of AI. It enables computers to comprehend, interpret, and produce human language. In design, NLP opens up a range of possibilities that enhance user interactions and streamline communication.

2.2.1 Conversational AI and Chatbots

One of the most prominent applications of NLP is **conversational AI**, particularly in the form of chatbots. Modern chatbots have advanced significantly beyond the simplistic, often frustrating bots of the past. Advanced NLP techniques now support sophisticated conversational interfaces that can handle complex user queries, deliver personalized assistance, and adapt their responses based on context and prior interactions.

These advancements significantly impact customer service, technical support, and interactive product experiences. The challenge for designers is to create conversational flows that feel natural and intuitive while balancing the power of AI with essential human elements, such as empathy and understanding. A well-designed chatbot aligns seamlessly with the brand's personality and integrates into the broader user experience. Moving beyond basic keyword matching, designers must now consider nuanced aspects such as tone, sentiment analysis, and the overall conversational journey.

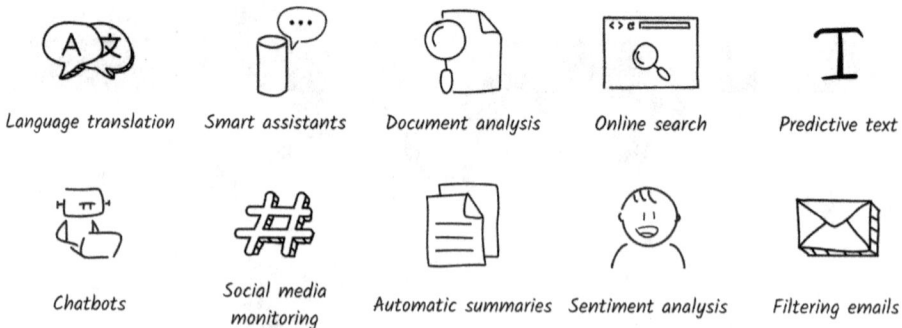

Language translation   Smart assistants   Document analysis   Online search   Predictive text

Chatbots   Social media monitoring   Automatic summaries   Sentiment analysis   Filtering emails

FIGURE 2.6   Ten applications of natural language processing.

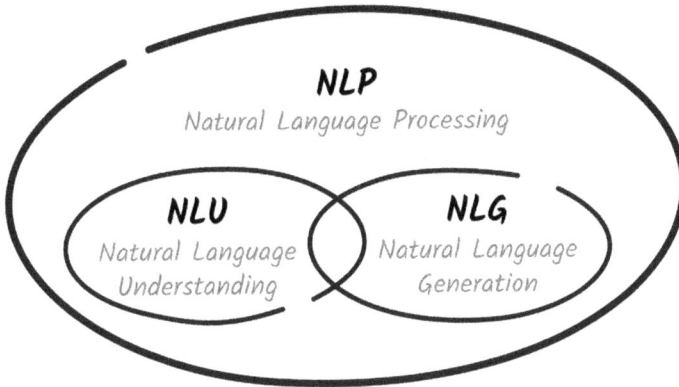

FIGURE 2.7   A Venn diagram illustrating the relationship between NLP, NLU, and NLG.

2.2.2 Natural Language Understanding

A key component of NLP is **natural language understanding** (**NLU**), which focuses on the syntactic and semantic analysis of text and speech to extract meaning. One essential capability of NLU is **named entity recognition**, which identifies important entities within the text, such as names, locations, and dates.

NLU uses syntactic and semantic analysis of text and speech to determine the meaning of a sentence. Consider the following sentences:

- "The **bank** is very comfortable."
- "I went to the **bank** to withdraw money."

While both sentences contain the word "bank," its meaning differs entirely based on context. In the first sentence, *bank* refers to a seating structure, whereas in the second, it denotes a financial institution. An effective NLU system must interpret these distinctions by analyzing contextual clues, surrounding words, and sentence structure to interpret them accurately. This contextual awareness makes NLU a powerful tool, enabling AI to process human language with greater accuracy and sophistication.

2.2.3 Natural Language Generation

While NLU enhances a computer's comprehension abilities, **natural language generation (NLG)** focuses on producing human-like text from structured data inputs. NLG applications must adhere to language rules governing morphology, lexicon, syntax, and semantics to generate coherent and contextually appropriate responses.

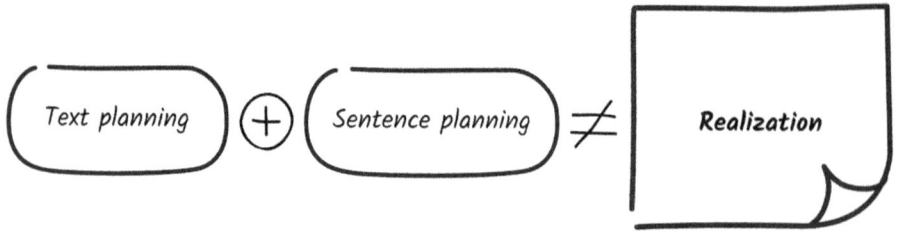

FIGURE 2.8 Beyond the sum of its parts: text planning and sentence planning are crucial steps, but they don't directly equate to final text realization.

NLG typically follows a three-stage process:

1. **Text planning**: Determines the logical structure and content order.
2. **Sentence planning**: Organizes punctuation, flow, and sentence construction.
3. **Realization**: Ensures grammatical accuracy and fluency in the final output.

By following this structure, NLG enables AI-powered systems to generate reports, automate responses, and create dynamic content tailored to users' needs.

Beyond conversational AI, NLP plays a crucial role in **document intelligence**, enabling experts to process text-based information and facilitate informed decision-making. Powered by ML, NLP and its subsets—NLU and NLG—are transforming industries ranging from healthcare and finance to customer support and legal services. As NLP technologies continue to evolve, their impact on human–computer interaction will only deepen, making them essential tools for designing more intuitive and intelligent digital experiences.

2.2.4 Knowledge Mining

Knowledge mining is a specialized subset of data mining that focuses on extracting valuable information from large volumes of often unstructured data to create a searchable knowledge repository. Although the terms "data mining" and "knowledge mining" are frequently used interchangeably, there are significant distinctions between the two concepts. Understanding these differences is essential for effectively leveraging both techniques.

2.2.4.1 Data Mining Versus Knowledge Mining

At a high level, **data mining** primarily focuses on uncovering hidden patterns, relationships, and correlations within large datasets. Its main goal is to transform raw data into actionable insights that inform decision-making processes. It emphasizes discovering trends and structures that may not be immediately evident.

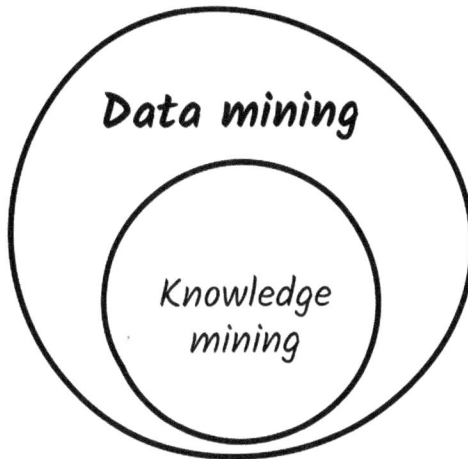

FIGURE 2.9   Knowledge mining as a specialized, outcome-driven subset of data mining.

In contrast, **knowledge mining** goes beyond identifying patterns; it focuses on extracting explicit, practical knowledge from data. It seeks to derive new insights and actionable knowledge that can be directly applied to solve problems or drive business strategies. Knowledge mining is not just about finding patterns but also about contextualizing and interpreting them to generate new understanding and strategic value.

2.2.4.2 Key Differences

- **Data mining**: Primarily concerned with discovering hidden patterns and relationships in data to support decision-making.
- **Knowledge mining**: Focuses on transforming patterns into actionable, practical knowledge that drives innovation and informs strategy.

In essence, data mining is about discovering what is hidden in the data, while knowledge mining is about transforming those discoveries into new, actionable knowledge. Consider an example from an e-commerce business that has collected a large dataset on sales over the past month.

Data mining: This technique would analyze shopping cart data to identify products that are frequently purchased together. It would uncover relationships between different items, such as customers often purchasing products A and B.

Knowledge mining: Once these patterns are identified, knowledge mining enables deeper analysis to extract valuable insights, such as how these associations can be used to optimize cross-selling and upselling strategies. By understanding the relationships between products, the business can personalize recommendations for individual customers, improving their shopping experience and ultimately increasing the average order value.

While data mining uncovers hidden patterns in data, knowledge mining transforms these insights into actionable knowledge that can significantly impact business strategy. The key distinction lies in the intent: data mining focuses on discovery, whereas knowledge mining emphasizes applying those discoveries in meaningful ways.

2.2.5 Generative AI

Generative AI refers to a category of capabilities designed to create original content. This technology has become increasingly prevalent, finding applications across various platforms, especially in chat-based interfaces. Generative AI models process natural language inputs and generate corresponding outputs in multiple formats, including text, images, code, and audio. Some of the most recognized tools in this area include **ChatGPT** for text generation, **MidJourney** for image creation, and **Synthesia** for video production.

2.3 DETECTION, PREDICTION, AND GENERATION

ML algorithms can be categorized into three primary applications: detection, prediction, and generation. Detection involves interpreting the present, while prediction entails outlining potential future paths. Additionally, machines are capable of performing generative or "creative" tasks.

The combination of detection and prediction opens the door to transformative innovation. Take autonomous vehicles, for example; systems must detect traffic signals, other vehicles, pedestrians, and obstacles in real time. But detection alone isn't enough—they must also predict how these elements will behave in the next few seconds. This seamless integration allows vehicles to operate safely and efficiently without human input, moving us closer to a future of intelligent, efficient, and responsible mobility.

2.3.1 Machine Learning Versus Predictive Analytics

Before concluding this chapter, it is important to clarify the distinction between **ML** and **predictive analytics (PA)**, as these terms are often used interchangeably but refer to distinct concepts.

Detection

Text and speech

Image interpretation

Human behavior and identity

Abuse and fraud

Prediction

Recommendations

Individual behavior and condition

Collective behavior

Generation

Visual art

Music

Text

Design

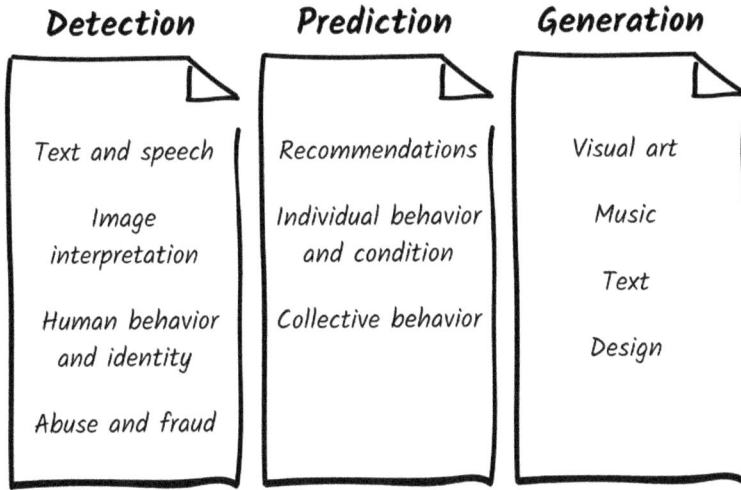

FIGURE 2.10   Three main categories of ML applications and their everyday use cases.

ML's primary purpose is to focus on developing adaptive models that can identify complex patterns within large datasets. These models learn from data over time and continuously improve their predictions as they are exposed to more information. ML enables systems to learn autonomously without requiring explicit programming for every new scenario.

On the other hand, **PA** is about leveraging statistical techniques to make specific predictions about future events based on historical data. While ML can be a tool within PA, PA often focuses on answering business-specific questions by identifying trends and patterns in past data. Its goal is to guide strategic decisions by anticipating future outcomes.

ML process involves developing sophisticated algorithms that learn and adapt from data. ML is an iterative process that generates new models from the data it processes. It often requires substantial computational resources and data to train models that can handle dynamic, unstructured problems.

PA process is typically business-driven, utilizing predefined statistical models or ML techniques to answer specific questions. For example, PA may analyze customer behavior patterns to predict future actions, such as which products they will likely purchase next. The focus is on delivering insights that can directly inform decision-making within an organization.

ML encompasses various techniques, ranging from simpler regression models to highly complex algorithms, such as neural networks and DL. These techniques often require an advanced understanding of mathematics, programming, and large datasets for training.

While **PA** can incorporate ML models, it is typically closer to advanced statistics and more predefined methods. It often relies on tools such as time series analysis or regression techniques, which are not always ML-based. PA tools are generally more accessible to business professionals and may not require extensive technical expertise in AI or algorithms.

ML is particularly suited for dynamic problems involving unstructured data, such as image or speech recognition. Its ability to learn and adapt makes it highly effective in fields where traditional rule-based systems would struggle, such as self-driving cars or NLP.

PA excels in strategic decision-making by providing insights based on historical trends and data. It is ideal for situations such as sales forecasting, risk assessment, and demand planning, where understanding past patterns can lead to more accurate predictions about future outcomes.

In summary, ML focuses on creating adaptive models that can learn from data and address complex, evolving problems. In contrast, PA utilizes historical data to make informed predictions that guide decision-making. ML is a powerful tool within PA; however, PA encompasses a broader range of techniques for forecasting and strategic insight.

TL;DR

This chapter expands on the foundational knowledge of artificial intelligence (AI) by showcasing core capabilities in action. It emphasizes that **not all AI is machine learning (ML)** and unpacks a variety of patterns—from visual perception to language understanding and content generation—that are redefining the boundaries of human–computer interaction.

AI ≠ just machine learning—While ML powers much of modern AI, the field also encompasses non-ML techniques and broader capabilities, including rule-based logic, vision, and language understanding.

1. Computer vision, giving machines sight
Enables systems to understand and act on visual input.

Core patterns:

- Image classification (e.g., Pinterest recommendations)
- Object detection (e.g., Amazon Go's checkout-free shopping)
- Semantic segmentation (e.g., Photoshop's "Select Subject" feature)
- Face recognition (e.g., Face ID)
- Optical character recognition (OCR) (e.g., Adobe Acrobat Pro's text extraction)

Real-world impact: Used in traffic systems, healthcare imaging, retail, and assistive tools like Microsoft Seeing AI.

Design note: Data bias is a serious concern; training datasets must be diverse and ethical to ensure fairness.

2. Natural language processing (NLP)
Allows machines to understand and generate human language.

Key subfields:

- **Natural language understanding** (**NLU**): Understands meaning and context (e.g., distinguishing "bank" as a seat versus a financial institution)
- **Natural language generation** (**NLG**): Generates human-like responses (e.g., automated reporting or ChatGPT)

Conversational AI: Sophisticated chatbots and virtual assistants now blend empathy, tone, and context for more human-centered dialogue.

3. Knowledge mining, from patterns to insights
There is a difference between data mining and knowledge mining:
- **Data mining** finds patterns in data, such as products that are frequently purchased together.
- **Knowledge mining** turns patterns into strategic insights, such as identifying cross-selling opportunities for frequently bought products.
- **Use case**: E-commerce businesses optimizing recommendation engines for personalization and upselling.

4. Generative AI, machines that create
- **Capabilities**: Capabilities include generating original text with tools like ChatGPT, creating images with MidJourney, and producing video content with Synthesia.
- **Power in design**: It facilitates creativity, iteration, and prototyping and empowers human creativity rather than replacing it.

5. Detection, prediction, and generation
Today, ML applications can be categorized into three buckets:
- **Detection** helps machines understand the present by identifying patterns and elements in real-time data.
- **Prediction** enables systems to anticipate future outcomes based on historical trends and user behavior.
- **Generation** enables AI to create original content such as text, images, or audio from learned data patterns.

What's the difference between machine learning and predictive analytics?

ML: Adaptive, data-driven, and autonomous. Used for complex and dynamic problems.
PA: Business-driven, structured forecasting, and often powered by ML, but not always.

Use cases:

- **ML**: Natural language, image recognition, and self-driving cars.
- **PA**: Forecasting sales, customer behavior, and risk management.

By understanding these core capabilities, designers and technologists alike can better harness AI's potential, not just to optimize interactions but to co-create the future of human-centered, intelligent systems.

Seven Fundamental AI Patterns

3

This chapter presents a practical framework for understanding AI through seven key patterns. We will examine each pattern in detail, demonstrating how they tackle real-world industry challenges. By recognizing these patterns, you will learn to spot opportunities for innovation and apply AI strategically to develop effective, user-centric solutions.

Imagine designing a new product that has the potential to revolutionize how people interact with technology. As you dive deeper into the project, you begin to realize that AI could play a pivotal role in making it smarter, more intuitive, and personalized. But how can you determine where to start or what to focus on? This is where recognizing AI patterns becomes crucial.

Think of AI patterns as a framework—an essential structure that helps you navigate the complex world of intelligent systems. These patterns allow you to spot opportunities and design functional solutions that are deeply connected to your users' needs. Grasping these patterns facilitates the creation of seamless, natural experiences that resonate with those interacting with your solutions.

This chapter will explore seven fundamental AI patterns: hyperpersonalization, recognition, patterns and anomalies, conversation and human interaction, PA and decisions, goal-driven systems, and, of course, the purpose of this book, autonomous and anticipatory systems. Each pattern represents a unique way AI can be exploited to solve problems and create value. From tailoring experiences to individual preferences with hyperpersonalization to enabling systems that operate independently with autonomy, these patterns provide a clear framework for designing smarter, more intuitive solutions. As we delve into each one, we will explore how they are applied to address challenges, enhance user experiences, and drive innovation across various industries.

DOI: 10.1201/9781003642800-4

3.1 HYPERPERSONALIZATION

Imagine a world where every digital interaction feels as if it were crafted just for you—your favorite music plays before you even search for it, your shopping cart fills with perfect recommendations, and your healthcare app suggests lifestyle changes tailored to your unique habits. This is not the future; it's the present, powered by hyperpersonalization. AI no longer just responds to user needs—it anticipates them. But with great customization comes great responsibility. The big question is how we balance precision with privacy. How can we ensure that hyperpersonalization enhances, rather than limits, the user experience? Let's explore how this AI pattern shapes industries and what it means for designers and decision-makers.

3.1.1 Personalization Versus Hyperpersonalization

At this point, you might wonder: what's the difference between personalization and hyperpersonalization? Traditional personalization tailors' experiences based on basic user data, such as preferences, purchase history, or demographics. Hyperpersonalization takes it a step further by leveraging real-time behavioral data, context, and PA to create highly dynamic and adaptive user experiences. Instead of offering broad recommendations based on past behavior, hyperpersonalization continuously learns and refines its understanding of the individual, allowing AI to:

- Display personalized content and recommendations (e.g., products, movies, news)
- Provide personalized guidance and insights (e.g., health tracking, financial planning)
- Adapt user interfaces and experiences dynamically
- Automate tailored plans and workflows according to user behavior

3.1.2 Real-World Applications

Hyperpersonalization is already reshaping multiple sectors:

- **Marketing and advertising**: AI-driven campaigns dynamically tailor promotions based on user interests and real-time behavior.
- **Finance and banking**: Personalized financial insights and fraud detection systems that adapt to individual transaction patterns.
- **Healthcare and wellness**: AI-powered recommendations for treatment plans, fitness regimens, and lifestyle adjustments.
- **Entertainment**: Streaming platforms like Netflix and Spotify refine recommendations based on evolving user preferences.
- **Education**: AI-assisted learning platforms that adapt content delivery to student progress and engagement levels.

According to Gartner, organizations that excel in personalization will outperform their competitors by 20%—a testament to the growing influence of this approach [6].

3.1.2.1 Case Study: Louis Vuitton's Hyperpersonalized Shopping Experience

When Louis Vuitton set out to elevate its e-commerce experience, it asked a bold question: why limit personalization to recommended products alone? The company realized that shoppers respond not only to *what* they see but also to *how* they perceive it—by leveraging Salesforce Einstein Designer, Louis Vuitton applied AI-driven hyperpersonalization to tailor the entire browsing experience, adapting page layouts, imagery, and style based on individual user behavior and personal preferences. This innovative approach highlighted the profound impact of hyperpersonalization, extending beyond mere product suggestions, and emphasized the critical role of design systems in this new era.

3.1.3 The Double-Edged Sword: Benefits and Risks

While hyperpersonalization unlocks new levels of user engagement and conversion, it comes with challenges.

Benefits:

- **Enhanced accessibility**: Customized experiences that adapt to user needs.
- **Greater relevance**: Reducing friction and delivering precisely what the user seeks.

FIGURE 3.1   Example of Louis Vuitton hyperpersonalization enabled by Salesforce Einstein [7].

- **Deeper engagement**: Strengthening user connection through highly personalized interactions.
- **Actionable insights**: AI-driven predictions that enhance business strategies.

Challenges and risks:

- **Data privacy concerns**: Continuous data collection raises ethical and security questions.
- **Filter bubbles and misinformation**: Over-personalization can trap users in echo chambers.
- **Misinterpretation of user data**: AI predictions are not always accurate, leading to flawed experiences.
- **Inconsistent recommendations**: Poorly optimized models can lead to irrelevant suggestions, which can damage user trust.
- **Overdependence on AI**: Users may rely too heavily on AI-driven recommendations, thereby diminishing their critical thinking and autonomy in decision-making.

3.1.4 Overcoming the "Cold-Start" Problem

Hyperpersonalization relies on data, but what occurs when there is very little? This is referred to as the **cold-start problem**, a challenge where AI systems struggle to deliver meaningful recommendations due to a lack of user data. Designers and engineers can address this by incorporating fallback strategies through:

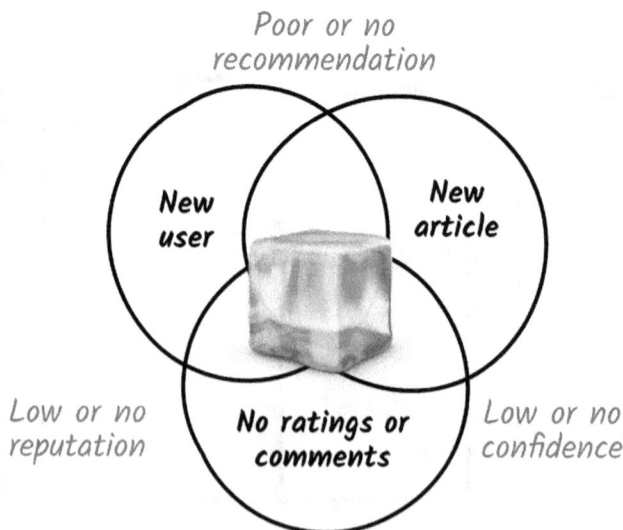

Poor or no recommendation

New user

New article

Low or no reputation

No ratings or comments

Low or no confidence

FIGURE 3.2 The three factors of the cold-start problem in hyperpersonalization [8].

- **Gradual profiling**: Allowing users to input initial preferences.
- **Hybrid models**: Combining AI-driven insights with traditional heuristics.
- **Transfer learning**: Leveraging insights from similar users or contexts.
- **Encouraging participation**: Designing experiences that incentivize data-sharing transparently.

A well-designed hyperpersonalization system anticipates this challenge, ensuring that users receive value even in their initial interactions. However, as AI plays a more active role in shaping user experiences, the question of accountability becomes critical. Who is responsible when hyperpersonalized systems misfire, exclude, or misinterpret user needs?

3.1.5 Accountability in AI

AI-driven hyperpersonalization isn't just about fine-tuning experiences—it's about responsibility. With AI making decisions that impact individuals, accountability must be embedded in the design and deployment process.

Accountability isn't about assigning blame—it's about promoting transparency, fairness, and continuous oversight [9]. Here are some key points to follow:

- **Precise user control**: Giving individuals transparency over how their data is used.
- **Bias mitigation**: Ensuring AI models don't reinforce harmful stereotypes or exclusions.
- **Redress mechanisms**: Creating pathways for users to challenge or correct AI-driven decisions.

Hyperpersonalization is a powerful tool; however, its ethical implications require careful consideration. The responsibility for ethical AI development does not rest solely on developers; it is a shared commitment among designers, product teams, and organizations. This journey necessitates ongoing dialogue, collaboration, and a dedication to ethical principles. The future of AI depends on it.

Only by embracing collective responsibility can we hope to navigate the complex ethical challenges posed by AI and build a future where this powerful technology serves humanity justly and equitably. Developing effective mechanisms and system feedback for redress and accountability is critical for both individual users and for overall trust and acceptance of AI technologies within society. Without these mechanisms, the potential benefits of AI may be overshadowed by valid concerns about unfairness, discrimination, and lack of recourse.

3.1.6 The Future of Hyperpersonalization

Hyperpersonalization has already transformed the way we interact with technology, and its potential continues to unfold. As AI evolves, its ability to create increasingly intuitive and meaningful experiences will also grow. However, the success of hyperpersonalization isn't solely about advancing AI's capabilities; it's about ensuring that those capabilities are utilized responsibly, transparently, and ethically.

As designers, strategists, and AI practitioners, our role is clear: to create hyperpersonalized experiences that empower users while protecting their privacy and trust. The future of AI isn't just about personalization—it's about doing personalization right.

3.2 RECOGNITION

Imagine walking into a store where the staff greets you by name and instantly understands your preferences. They recall what you purchased last time, suggest items that enhance your style, and even anticipate your needs before you express them. Now, translate this seamless, intuitive interaction into the digital realm—this is the power of AI-driven recognition.

Recognition is one of the most fundamental AI patterns, enabling machines to perceive, classify, and respond to information in a manner similar to humans. From facial recognition that unlocks your phone to AI systems detecting fraudulent transactions in real time, recognition-based AI enhances convenience, security, and efficiency across various industries. However, as with any powerful tool, its applications raise important questions about privacy, bias, and ethical responsibility. Let's explore how recognition shapes the digital experiences we engage with daily and what designers need to consider when creating AI-driven solutions.

3.2.1 Types of AI-Driven Recognition

AI-driven recognition spans a broad spectrum of applications, each leveraging ML to process and interpret different types of input:

- **Image and facial recognition**: Used in security systems, social media tagging, and biometric authentication.
- **Speech and voice** recognition enable virtual assistants like Alexa and Siri, facilitate real-time transcription, and support customer service automation.
- **Text recognition (OCR and NLP)**: Transforming handwritten or printed text into digital formats and facilitating AI-driven language understanding.
- **Pattern recognition**: Detecting anomalies in financial transactions, cybersecurity threats, or medical imaging.
- **Behavioral recognition**: Analyzing user interactions to detect fraud, assess mood, or personalize experiences.

These capabilities make recognition a cornerstone of AI-driven innovation, enhancing how humans and technology interact meaningfully.

3.2.2 Real-World Applications

Recognition technology is already embedded in many industries:

- **Security and authentication**: Face unlock systems, fingerprint scans, and fraud detection prevent unauthorized access.
- **Retail and e-commerce**: Visual search engines allow customers to shop by uploading images instead of typing queries.
- **Healthcare**: AI analyzes medical scans to detect diseases earlier and more accurately than human doctors.
- **Customer support**: AI-powered chatbots and virtual assistants recognize voice patterns and sentiment to provide more human-like interactions.
- **Transportation and smart cities**: License plate recognition for automated toll payments, traffic monitoring, and enhanced public safety.

3.2.2.1 Case Study: Apple Face ID—A Seamless but Controversial Innovation

When Apple introduced Face ID, it revolutionized smartphone security by making biometric authentication seamless and nearly instant. However, its rollout also sparked debates about privacy and bias. Face ID struggled with accuracy in recognizing certain ethnic groups, exposing the risk of biased AI models. Apple had to refine its algorithm to ensure fairer and more inclusive recognition. This case highlights the dual nature of recognition technology—offering convenience while demanding careful consideration of fairness and ethics.

3.2.3 The Double-Edged Sword: Benefits and Risks

While recognition enhances security and personalization, it also raises concerns:

Benefits:

- **Enhanced security**: Stronger authentication methods reduce fraud and unauthorized access.
- **Efficiency and automation**: AI accelerates processes that would take humans much longer to perform manually.
- **Improved user experience**: Recognition enables frictionless interactions and personalized services.
- **Advanced diagnostics**: AI can detect diseases, errors, or anomalies with high accuracy, assisting human decision-making.

Challenges and risks:

- **Data privacy concerns**: Constantly tracking users raises ethical and legal issues about consent and surveillance.

- **Algorithmic bias**: If trained on biased datasets, recognition systems may disproportionately fail for specific demographics.
- **False positives and errors**: Misidentifications can have severe consequences, from wrongful arrests to medical misdiagnoses.
- **Overreliance on AI**: Automated decisions based on recognition may remove human oversight from critical processes.

3.2.4 Mitigating Recognition Bias and Errors

Bias represents one of the key challenges in recognition systems. When an AI model is primarily trained on a particular demographic, its accuracy can decline for groups that are less represented. To address this, the following steps are necessary.

- **Diverse and representative training data**: Ensuring datasets include a broad spectrum of inputs to minimize bias.
- **Transparency in AI decision-making**: Making AI logic interpretable to users so they can understand why a system made a specific decision.
- **Human oversight**: Combining AI insights with human judgment to mitigate errors and unintended consequences.

3.2.4.1 Accountability in Recognition-Based AI

As AI recognition systems gain influence in decision-making, accountability becomes crucial. Who is responsible when a facial recognition system falsely identifies an individual? How can users contest AI-driven outcomes? These questions highlight the need for the following:

- **Regulatory frameworks**: Governments and organizations must establish policies that govern the ethical use of AI.
- **User control and consent**: Users should have clear options to manage their data and opt out of AI-based recognition when appropriate.
- **AI explainability**: Developers should design AI models that justify their outputs, reducing the "black box" effect.

3.2.5 The Future of Recognition AI

Recognition transforms our digital interactions, from securing devices to enabling smarter services. However, its success depends on technological advancements and the ethical and responsible implementation. As AI continues to evolve, our strategies for ensuring that recognition-driven innovations are fair, transparent, and accountable must also evolve.

As designers and AI practitioners, we aim to harness the recognition of AI's potential while safeguarding against its risks. The future of recognition is not just about teaching machines to "see" and "understand"—it's about ensuring that they do so in ways that enhance, rather than compromise, human dignity and trust.

3.3 PATTERNS AND ANOMALIES

Recognizing patterns and detecting anomalies is critical for decision-making in a data-driven world. AI systems trained to identify recurring trends, and unexpected deviations enable industries to optimize processes, enhance security, and even predict future events. Whether flagging fraudulent transactions, detecting cyber threats, or identifying early signs of disease, AI-driven pattern recognition plays a pivotal role in making sense of complex information.

Pattern recognition forms the foundation of AI's ability to interpret data. It enables machines to analyze vast datasets, uncover relationships, and predict likely outcomes. However, what occurs when data deviates from the expected? This is where anomaly detection comes into play—AI can identify outliers that may indicate fraud, malfunctions, or rare yet critical events.

3.3.1 How AI Recognizes Patterns and Anomalies?

AI identifies patterns by learning from historical data, uncovering correlations, and applying statistical and ML techniques to categorize information. Similarly, it detects anomalies by recognizing irregularities that significantly deviate from the norm.

- **PA**: AI forecasts trends based on historical patterns, such as stock market predictions and demand forecasting.
- **Fraud detection**: Identifies suspicious activities in banking and e-commerce transactions by flagging deviations from typical user behavior.
- **Cybersecurity**: Detects unusual login attempts, unauthorized access, or malware activities.
- **Healthcare diagnostics**: Identifies irregularities in medical scans or vital signs to detect potential health risks early.
- **Manufacturing and IoT**: Predicts equipment failures by spotting deviations in sensor data.

3.3.1.1 Real-World Applications

Pattern and anomaly detection are already transforming various industries:

- **Finance and fraud prevention**: Banks use AI to detect fraudulent transactions by comparing user behavior against expected norms.
- **Healthcare and medical imaging**: AI analyzes medical scans to identify disease-related patterns and flag anomalies for further examination.
- **Retail and consumer behavior**: Analyzes shopping patterns to enhance inventory management, pricing strategies, and personalized recommendations.

- **Cybersecurity**: AI monitors network traffic and detects anomalies that may indicate a data breach or hacking attempt.
- **Autonomous vehicles**: AI promotes safety by recognizing predictable driving behaviors while flagging unexpected obstacles or erratic driving patterns.

3.3.1.2 Case Study: Zebra Medical Vision's AI-Powered Diagnostic Tool

Zebra Medical Vision is a platform that uses computer vision techniques combined with ML to analyze medical images, including X-rays, CT scans, and MRIs. The company developed algorithms capable of detecting anomalies in medical images, helping healthcare professionals diagnose a wide range of conditions, including cancer, heart disease, and lung disease. This AI-driven approach significantly enhances diagnostic accuracy, providing faster and more reliable results that enable doctors to make more informed decisions.

3.3.2 The Double-Edged Sword: Benefits and Risks

While AI-driven pattern recognition offers immense value, it also presents challenges that require careful consideration.

Benefits:

- **Enhance efficiency**: Automates complex tasks, allowing faster and more accurate decision-making.

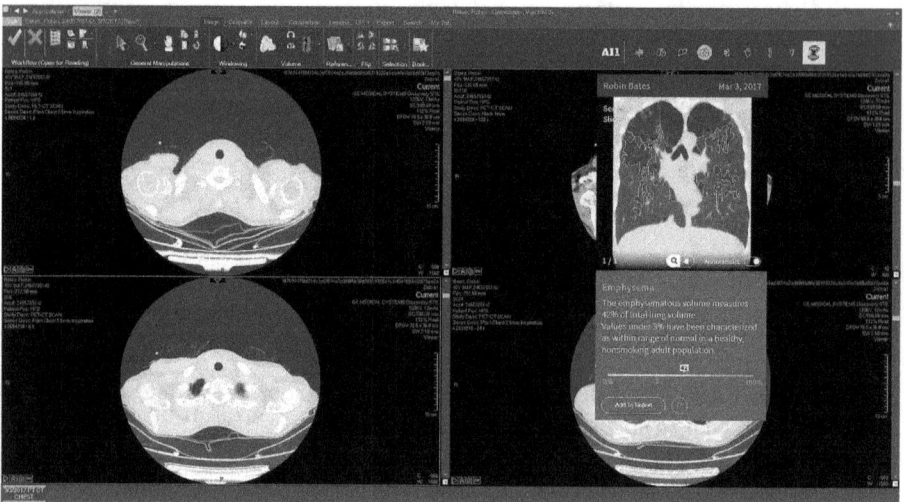

FIGURE 3.3   Zebra Medical Vision's AI-powered diagnostic tool [10].

- **Improved security**: Detects threats and fraud in real time, preventing financial and cyber risks.
- **Better resource management**: Optimizes inventory, energy consumption, and predictive maintenance.
- **Early problem detection**: Identifies issues before they escalate, whether in healthcare, finance, or industrial systems.

Challenges and risks:

- **False positives and negatives**: AI may misclassify anomalies, leading to unnecessary actions or missed threats.
- **Bias in training data**: If AI is trained on biased datasets, it may overlook critical anomalies or reinforce existing prejudices.
- **Complexity in interpretation**: Some AI models function as "black boxes," making it difficult to understand why specific patterns or anomalies were identified.
- **Privacy concerns**: Continuous monitoring and data analysis raise ethical considerations about user consent and data security.

3.3.3 Biased Data in Predictive Analytics and Decisions

PA relies heavily on historical data; however, AI can reinforce biases present in that data. If a bank's loan approval AI is trained on past approvals that favored specific demographics, it may **unfairly reject applications from marginalized groups**. Similarly, biased hiring algorithms can perpetuate discrimination by prioritizing candidates who resemble those who have been hired in the past. To address this issue, designers must question engineers about the following:

- **A diverse training dataset** is available that represents a wide range of backgrounds and behaviors.
- **They use fairness-aware ML techniques** to detect and correct biased patterns.
- **Continuously audit AI models** to identify and mitigate discrimination in decision-making.
- **Design mechanisms to indicate when a variable or user may be at risk** of failing in an outlier range.

3.3.3.1 Ensuring Responsible AI in Pattern and Anomaly Detection

To maximize the benefits and mitigate risks, organizations must:

- **Use diverse and unbiased datasets**: Ensure AI models are trained on representative data to minimize errors and biases.
- **Implement explainable AI (XAI)**: Develop models that provide transparent reasoning behind their pattern recognition.

- **Maintain human oversight**: AI should augment, not replace, human judgment in critical decision-making processes.
- **Strengthen privacy measures**: Securely handle user data and ensure compliance with regulations such as the European general data protection regulation (GDPR).

3.3.4 The Future of Pattern and Anomaly Recognition

AI's ability to recognize patterns and detect anomalies is revolutionizing industries by enabling more intelligent decision-making and proactive risk management. However, the technology must be deployed responsibly to ensure accuracy, fairness, and the ethical use of data.

As AI practitioners and designers, our challenge is to develop robust recognition systems and improve them to be fair, interpretable, and trustworthy. The future of pattern recognition AI is not solely about identifying what is normal and abnormal; it is also about ensuring that these definitions serve everyone equitably and responsibly.

3.4 CONVERSATION AND HUMAN INTERACTION

Imagine a world where you can chat with a healthcare professional—not in person, but through an AI-driven virtual assistant. This assistant understands your queries, interprets your needs, and provides tailored responses just like a human would. However, there's no person on the other side—only a machine powered by advanced algorithms. This is the essence of AI in conversation and human interaction. AI-powered systems have become increasingly adept at understanding, processing, and responding to human language, making them indispensable tools in various industries, from customer service to healthcare.

The key challenge with conversational AI is creating systems that feel human-like without venturing into uncomfortable territory. These systems, known as conversational agents, utilize a blend of NLP, ML, and sometimes DL to interpret and generate responses. Let's explore the applications, real-world examples, benefits, and challenges of these AI patterns.

3.4.1 What This Patterns of AI Allows Us to Do?

- **Automated customer service**: Chatbots and virtual assistants are capable of answering customer inquiries and resolving issues instantly, eliminating the need for human intervention.

- **Voice interaction**: Voice assistants, such as Siri, Alexa, and Google Assistant, can comprehend spoken commands and execute tasks accordingly.
- **Educational support systems**: Virtual tutors that help students by offering immediate feedback, guidance, and tailored learning experiences.
- **Interaction in digital media**: Bots that engage users on social media platforms through real-time conversations, offering information, or managing simple tasks.

3.4.2 Real-World Applications

AI-driven conversation systems are transforming various industries, and their applications are extensive:

- **Customer support**: AI-powered chatbots and virtual assistants manage customer inquiries, offer immediate answers, and resolve issues without requiring human intervention.
- **Healthcare**: Virtual health assistants assist patients with symptom checking, provide medical advice, and offer mental health support.
- **Finance**: Chatbots in banking and insurance assist customers in managing accounts, executing transactions, and resolving financial inquiries.
- **Retail**: AI assistants on e-commerce platforms guide users in purchasing decisions, respond to product-related questions, and monitor orders.
- **Smart Devices**: Voice-enabled assistants integrated into smart devices allow users to control their homes (lights, heating, security systems) by merely speaking commands.

3.4.2.1 Use Case: Supporting Mental Health Through Conversational AI

Wysa is a powerful example of how conversational AI can support well-being in deeply human ways. Designed as a mental health companion, Wysa utilizes an AI-powered chatbot to facilitate sensitive, empathetic conversations that help users manage stress, anxiety, and low mood. Rather than offering generic advice, Wysa uses techniques from cognitive behavioral therapy, dialectical behavior therapy, and other evidence-based practices to guide users through structured yet natural-feeling conversations.

The app begins with simple prompts, such as "How are you feeling today?" and gently follows up with personalized, therapeutic dialogue. Users can explore a range of tools—from mood tracking and guided journaling to breathing exercises and thought reframing—all delivered through conversational interaction. If deeper support is needed, Wysa offers access to human therapists, ensuring a seamless transition from AI assistance to professional care.

So far, Wysa has proven particularly effective in lowering the barrier to seeking help, especially in contexts where stigma, cost, or access to traditional mental health resources might prevent people from reaching out. It's an elegant example of conversation and human interaction pattern, not just because it *talks to* users but because it also truly *listens* and helps them move toward better mental health through conversation.

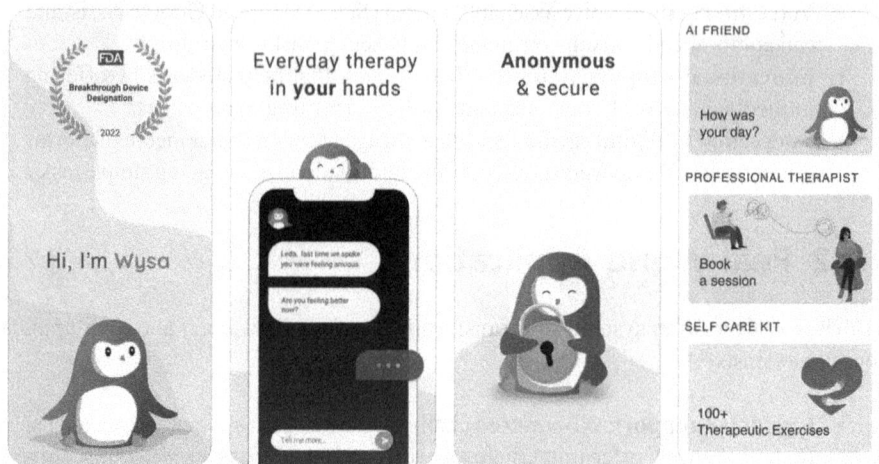

FIGURE 3.4   A 2025 screenshot of the Wysa Health service [11].

3.4.3 The Double-Edged Sword: Benefits and Risks

While this AI-driven pattern presents immense potential, it also poses challenges that require careful consideration.

Benefits:

- **24/7 availability**: AI systems never require sleep or breaks, allowing users to receive assistance at any time.
- **Cost-effective**: Reduces reliance on human agents for customer service and medical consultations, thereby lowering operational costs.
- **Scalability**: AI systems can manage a large number of interactions simultaneously, which would be unfeasible for humans alone.
- **Efficiency**: Delivers instant responses, speeding up processes like problem-solving, symptom assessment, or feedback on learning.
- **Personalization**: AI systems can adapt to user needs, tailoring conversations to individual preferences or previous interactions.

Challenges and risks:

- **Miscommunication**: AI continues to struggle with understanding nuanced language, humor, and emotional cues, which can lead to misunderstandings.
- **Privacy concerns**: AI systems frequently collect sensitive user data, which poses risks of potential data breaches or misuse.

FIGURE 3.5 Screenshot of IBDM Cat's Movie web page.

Source: www.imdb.com/title/tt569757

- **Overreliance on automation**: Excessive dependence on AI for essential tasks, like medical advice or customer support, may lead to missed opportunities for human judgment and oversight.
- **Bias**: If conversational AI is trained on biased data, it may provide skewed or unfair responses, which can negatively impact the user experience.

3.4.3.1 The Uncanny Valley Effect in Conversational AI

The Uncanny Valley effect is one of the most critical challenges for conversational AI. It refers to the discomfort or eerie feeling people experience when interacting with AI that is nearly human-like but still noticeably artificial. For example, a voice assistant or chatbot that sounds and behaves almost like a human but has subtle, unnatural quirks can create an unsettling user experience.

In the case of Wyse Health, if its AI-powered assistant were to mimic human-like conversations too closely but fall short in key areas, such as empathy or medical accuracy, users might feel uneasy about the interaction. This could lead to distrust, particularly in the context of a personal matter such as healthcare.

3.4.3.2 Mitigating the Uncanny Valley Effect

To avoid triggering the uncanny valley effect, designers and developers can take several steps:

- **Establish clear expectations**: Users should understand that they are engaging with an AI, so they should not expect the same level of empathy or emotional intelligence as they would from a human.

- **Human-like, but not human**: Design AI interactions to be friendly while avoiding excessive human-like qualities. A warm and approachable tone is preferred without closely mimicking human emotions or behaviors.
- **Transparent communication**: Ensure that the AI clearly articulates its limitations, such as indicating that it is an AI and not a licensed medical professional, as seen with Babylon Health. Transparency fosters trust.
- **Consistency and reliability**: Ensure that the AI's responses are accurate and consistent. When users receive trustworthy information, the discomfort caused by artificiality is minimized.

3.4.4 The Future of Conversational AI

Conversational AI is revolutionizing the way we interact with machines, providing instant, scalable, and personalized experiences across various industries. The potential is enormous, whether in healthcare, customer service, or education. However, as AI systems become more human-like, it's crucial to ensure that these technologies are designed thoughtfully to avoid uncomfortable experiences, such as those caused by the Uncanny Valley effect.

As AI continues to evolve, our role as designers and developers is to create systems that build trust, offer value, and prevent unsettling users. By doing this, we can ensure that conversational AI remains a valuable asset in shaping the future of human–computer interactions.

3.5 PREDICTIVE ANALYTICS AND DECISIONS

Earlier in Chapter 1, we examined the role of PA in shaping AI-driven decision-making, where AI analyzes historical data to anticipate future trends. This foresight enables businesses to make proactive decisions, such as adjusting inventory based on demand patterns or forecasting market shifts before they happen.

Now, let's elaborate on that concept by exploring how PA and decision-making systems function, their real-world applications, and the ways they can transform industries.

Imagine a retail company anticipating a surge in demand for winter coats before the first snowflake falls or a financial institution identifying potential loan defaults before they occur. This foresight isn't magic; it's the power of PA—AI systems that analyze historical data to forecast future outcomes, enabling businesses to make proactive and informed decisions. PA uses statistical models and ML algorithms to analyze past patterns and predict future events. Companies can optimize operations, enhance customer experiences, and mitigate risks by identifying key variables that influence outcomes.

3.5.1 How Predictive Analytics and Decisions Work?

- **Data collection and preparation**: Collect pertinent historical data from multiple sources, making sure it is clean and well organized for analysis.
- **Statistical modeling and ML**: Applying algorithms to uncover patterns and relationships within the data. Techniques include regression analysis, decision trees, and neural networks.
- **Identification of key variables**: Determining which factors have the most significant impact on desired outcomes, such as customer behavior or equipment failure.
- **Prediction generation**: Using the developed models to forecast future events or trends.
- **Automated decision systems**: Implementing AI-driven systems that utilize these predictions to recommend or carry out actions, frequently without human intervention, such as adjusting inventory levels or flagging fraudulent transactions.

3.5.2 Real-World Applications

PA is revolutionizing multiple industries by facilitating data-driven decision-making.

- **Retail**: Companies analyze customer purchase histories and browsing behaviors to predict demand, optimize inventory levels, and refine marketing strategies.
- **Finance**: Financial institutions employ predictive models to evaluate credit risk, identify fraudulent activities, and predict market trends.
- **Manufacturing**: Predictive maintenance models foresee equipment failures, enabling timely interventions and minimizing downtime.
- **Healthcare**: Analyzing patient data helps predict disease outbreaks, patient admissions, and treatment outcomes, facilitating proactive care.
- **Supply chain and logistics**: PA supports demand forecasting, optimizes routes, and manages inventory, improving efficiency and cutting costs.

3.5.2.1 Use Case: Lokad—Enhancing Supply Chain Decisions With Predictive Analytics

Lokad, a French software company, exemplifies specialization in supply chain optimization through PA. Lokad's platform leverages ML and probabilistic forecasting to analyze historical sales data and market trends, enabling businesses to anticipate demand fluctuations and optimize their inventory levels. This method helps companies avoid stockouts and overstock situations, leading to cost savings and enhanced customer satisfaction.

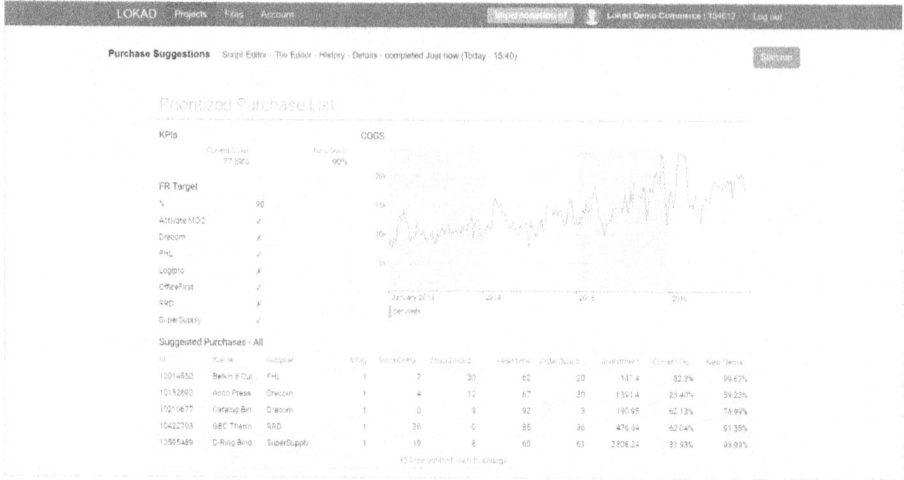

FIGURE 3.6   A screenshot from the Lokad supply chain application taken in 2025.

Source: Available at www.lokad.com/features

3.5.3 The Double-Edged Sword: Benefits and Risks

While this AI-driven pattern offers immense potential, it also presents challenges that require careful consideration.

Benefits:

- **Enhanced forecasting accuracy**: Improved demand and trend predictions lead to better planning and resource allocation.
- **Operational efficiency**: Optimizing processes based on predictive insights reduces waste and streamlines operations.
- **Risk mitigation**: Identifying potential issues before they occur enables proactive measures, thereby minimizing negative impacts.
- **Personalized customer experiences**: Tailoring offerings based on predictive insights enhances customer satisfaction and loyalty.
- **Competitive advantage**: Leveraging data-driven decisions enables businesses to stay ahead of market trends and competitors.

Challenges and risks:

- **Data quality and availability**: Accurate predictions rely on high-quality, comprehensive data; poor data can lead to unreliable outcomes.
- **Model complexity and interpretability**: Complex models can be challenging to comprehend and explain, resulting in a loss of trust and difficulties with regulatory compliance.

- **Overfitting**: When a model learns from data too precisely, including noise and exceptions, it may struggle to adapt to new, unseen situations. This reduces its ability to make accurate predictions in real-world scenarios.
- **Ethical concerns**: PA can lead to privacy issues and possible biases, particularly when sensitive personal data is involved.
- **Dependence on technology**: Overreliance on automated decisions may overlook human judgment and lead to unintended consequences.

3.5.4 The Future of Predictive Analytics and Decisions

For designers, PA represents an opportunity to create smarter and more intuitive experiences that anticipate user needs and behaviors. By leveraging historical data and forecasting tools, designers can optimize interfaces, enhance personalization, and streamline decision-making processes. However, it is crucial to approach predictive models with a critical eye. The accuracy of predictions heavily depends on data quality, and designers must collaborate with data scientists to ensure that AI models are transparent, ethical, and inclusive.

Additionally, acknowledging the limitations of these systems and planning for potential failures are essential for mitigating risks and building user trust. As AI continues to develop, designers must ensure that predictive systems enhance efficiency while prioritizing user experience, fairness, and accountability. By carefully incorporating PA into their work, designers can contribute to the creation of more responsive and impactful solutions that align with user needs and business objectives.

3.6 GOAL-DRIVEN SYSTEMS

Goal-driven systems in AI are designed with a clear, predefined objective. Unlike other AI patterns that might respond to inputs, goal-driven systems adopt a proactive approach. They continuously evaluate their environment, process data, and make decisions to achieve specific outcomes. Whether ensuring safety, improving efficiency, or personalizing user experiences, these systems adapt in real time based on new information, adjusting their actions to meet established goals.

Goal-driven systems in AI are designed to achieve specific, predefined objectives. Unlike other AI patterns that may react to inputs, these systems adopt a proactive approach. They continually assess situations and make decisions to achieve their goals, whether it is ensuring safety, improving efficiency, or providing a personalized experience. These systems adapt by learning from past actions and refining their strategies, often acting autonomously to achieve the desired results. This characteristic makes goal-driven AI particularly powerful for applications that demand long-term planning, complex decision-making, and ongoing optimization.

While goal-driven systems can employ various ML techniques, many, especially those that operate autonomously or in dynamic environments, tend to utilize DL methods. DL allows the system to process vast amounts of data and learn complex patterns without human intervention. It assists these systems in making decisions that evolve over time, refining their strategies to align more closely with their objectives.

3.6.1 The Role of Machine Learning and Deep Learning in Goal-Driven Systems

Goal-driven systems utilize ML and DL to continually adapt and make decisions based on the data they process. The choice between ML and DL typically depends on the tasks' complexity and the data volume the system must manage.

In ML, goal-driven systems utilize algorithms to process data, identify patterns, and make predictions or recommendations based on historical information. These systems learn from experience and adjust their strategies accordingly, which makes them particularly effective for applications such as PA, anomaly detection, and optimization tasks.

In DL, it becomes essential for more complex decision-making, especially when managing large amounts of unstructured data, such as images, speech, or extensive datasets. DL models, particularly neural networks, can process and learn from vast quantities of data to recognize patterns that would be challenging for traditional ML algorithms to identify. DL is often employed in goal-driven systems when the task involves high-level abstraction, such as autonomous decision-making, NLU, and intricate simulations.

By combining the strengths of both ML and DL, goal-driven systems can make highly informed decisions, optimize actions over time, and adapt to a wide range of dynamic environments. These systems continually refine their processes to meet their predefined goals for safety, efficiency, or personalized experiences.

3.6.2 What Goal-Driven Systems Enable?

Goal-driven AI enables the following:

- **Autonomous navigation**: Refers to self-driving cars capable of navigating complex environments while optimizing routes to their destinations.
- **Personalized experiences**: AI customizes recommendations, offers, and content according to user behavior and preferences.
- **Predictive maintenance**: Involves AI systems that anticipate failures in machinery or vehicles and take proactive measures to prevent them, aiming to minimize downtime.
- **Performance optimization**: AI in business and industry identifies key performance indicators (KPIs) and drives actions to enhance outcomes, such as increasing sales or boosting employee efficiency.

3.6.3 Real-World Applications

Goal-driven systems are increasingly present across different sectors:

- **Autonomous vehicles**: Self-driving cars navigate through traffic, optimize fuel efficiency, and reach their destinations while continuously adapting to real-time variables.
- **Personalized marketing**: AI systems in e-commerce platforms enhance product recommendations by analyzing user preferences and behavior to boost engagement.
- **Manufacturing and industrial operations**: AI systems in factories can oversee production lines, anticipate when machines will require maintenance, and guarantee that operations function at peak efficiency.
- **Healthcare**: Personalized health plans and AI-driven diagnostics designed to achieve health goals, including enhancing patient outcomes and minimizing hospital readmissions.

3.6.3.1 Use Case: Tesla's Autopilot—Driving Toward Autonomy

Tesla's Autopilot exemplifies a powerful goal-driven AI system in action. Designed for full autonomy, Tesla's self-driving technology continually learns from vast amounts of driving data to enhance its performance. The AI system utilizes real-time sensor inputs and ML algorithms to make decisions, such as adjusting speed, changing lanes, or applying the brakes—all while striving to reach the final destination as safely and efficiently as possible.

Tesla's goal-driven system continuously improves with each update, guided by the vision of enabling fully autonomous vehicles. While the system can operate

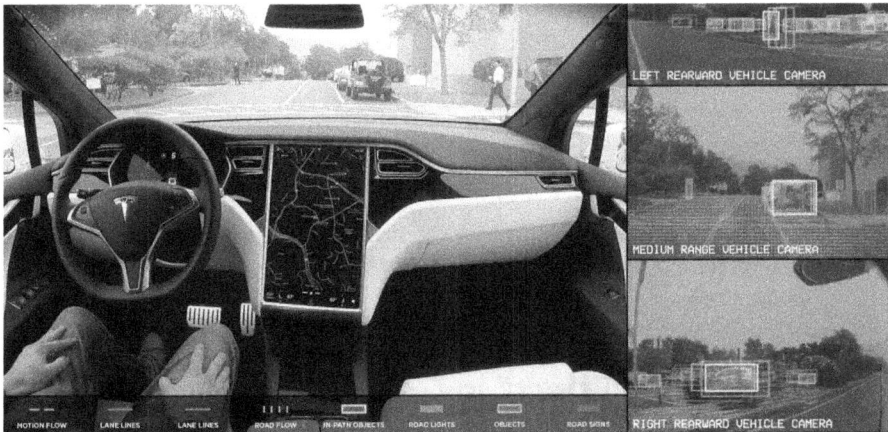

FIGURE 3.7   Tesla Autopilot self-driving service [12].

autonomously under certain conditions, the ultimate goal remains to have fully self-driving cars that can function independently of human intervention. This system evolves to meet new challenges, such as detecting unexpected obstacles and navigating complex environments, all in line with the goal of safe and efficient transportation.

3.6.4 The Double-Edged Sword: Benefits and Risks

While goal-driven AI offers immense possibilities, it also presents challenges that require careful planning and foresight.

Benefits:

- **Efficiency**: These systems are designed to optimize processes and achieve outcomes more effectively than manual methods, such as enhancing route efficiency in autonomous vehicles or automating customer support.
- **Autonomy**: Goal-driven systems can make decisions with minimal human intervention, facilitating smoother operations in fields such as transportation or healthcare.
- **Continuous improvement**: DL algorithms in goal-driven systems enable ongoing adaptation to new data and environments, allowing for ongoing improvement.
- **Personalization**: By concentrating on specific objectives, AI can provide highly personalized experiences.

Challenges and risks:

- **Ethical implications**: In decision-making systems, such as autonomous vehicles, AI must be programmed to make choices in complex, morally ambiguous situations. This raises questions regarding accountability and ethics.
- **Safety concerns**: Goal-driven systems must be highly reliable, especially in high-risk environments such as self-driving cars. A minor error in judgment can result in catastrophic consequences.
- **Lack of transparency**: Goal-driven systems often operate as "black boxes," making it challenging for users to understand the decisions made, which results in diminished trust.
- **Overfitting**: A goal-driven system that is overly focused on achieving a specific target may optimize excessively for that goal, resulting in unintended consequences or inefficiencies.

3.6.5 The "Halo Effect" in Goal-Driven Systems

Goal-driven AI systems often optimize for specific objectives. Still, **when a system excels in one area, users may mistakenly assume it is equally competent in** others—a cognitive bias known as the **Halo Effect**. For example, a navigation AI that accurately predicts traffic may create an illusion of infallibility, leading users to blindly trust its recommendations, even when human judgment is necessary. To mitigate the Halo Effect, designers should:

- **Clarify the capabilities and limitations of AI**: Utilize interface cues and messaging to convey when AI outputs are based on probabilities rather than certainties.
- **Encourage user verification**: Implement mechanisms that prompt users to reevaluate AI-driven recommendations in critical situations.
- **Avoid overpromising**: Ensure that branding, UI copy, and system behavior set realistic expectations about what AI can and cannot do.
- **Implement explainability features**: This enables users to understand why a particular decision was made, thereby reducing misplaced trust and confidence.

3.6.6 The Future of Goal-Driven AI

Goal-driven systems transform industries by enabling intelligent decision-making and optimizing processes for specific outcomes. Tesla's Autopilot, for example, demonstrates how AI can be leveraged to pursue a complex, long-term goal, such as full autonomy. As AI advances, goal-driven systems will increasingly integrate into our lives, providing smarter and more efficient solutions.

While these systems often rely on ML and DL to adapt continuously, ML typically handles less complex decision-making tasks, such as predicting outcomes based on historical data. At the same time, DL facilitates more advanced, high-level abstractions necessary for functions like autonomous driving or NLP. These technologies empower goal-driven systems to operate autonomously and make smarter, long-term decisions. However, as with all AI, careful consideration of ethical implications, transparency, and user safety will be key to their successful and responsible implementation.

As designers, we ensure that these systems align with human needs and values while promoting innovation and progress. The future of goal-driven AI lies in developing solutions that not only achieve objectives but also enhance trust, fairness, and user experience.

3.7 AUTONOMOUS AND ANTICIPATORY SYSTEMS

Lastly, we have autonomous and anticipatory systems that represent the forefront of AI. These machines operate independently, anticipating future needs, behaviors, and decisions. They go beyond merely reacting to user inputs; instead, they proactively predict and make decisions in real time, often before any explicit user request is made. This capability allows them to function seamlessly in environments where human oversight is minimal or impractical, and their actions seem driven by an inherent understanding of both the environment and user needs.

From self-driving cars that navigate roads and make decisions in complex environments to intelligent tutoring systems that anticipate students' learning needs, autonomous and anticipatory systems are designed to optimize outcomes even before the user realizes what they require.

This AI pattern will be the primary focus of this book. There is a significant gap in the existing literature regarding the effective addressing of challenges and risks involved in designing such systems. This chapter introduces the core principles, explores a real-world application, and highlights how this AI pattern reshapes industries and societies.

3.7.1 What Autonomous and Anticipatory Systems Can Do?

Autonomous and anticipatory systems extend beyond merely reacting to inputs—they predict future events and make decisions that shape what comes next. Here's what these systems can accomplish:

- **Autonomous decision-making**: These systems function independently and make decisions based on real-time data, such as an AI-powered drone navigating obstacles without human intervention.
- **Anticipating user needs**: By examining previous behaviors and preferences, these systems can predict what users will need before they explicitly request it. For instance, AI assistants may propose calendar events or offer shopping suggestions before you even consider them.
- **Adaptive learning**: These systems adjust to their environments, consistently enhancing their performance over time. They become more efficient when faced with new data and challenges, making them highly responsive and personalized.
- **Efficient resource allocation**: In logistics environments, these systems can forecast demand and adjust resources accordingly, such as rerouting delivery trucks to optimize fuel consumption or deploying maintenance staff before equipment fails.

3.7.2 Real-World Applications

Autonomous and anticipatory systems are already transforming various industries:

- **Education**: In education, autonomous and anticipatory systems can personalize learning experiences to meet individual needs. These systems analyze student performance and learning patterns in real time, adjusting content and pacing based on individual needs. This proactive approach ensures that learning gaps are addressed before they become obstacles, providing a tailored educational experience for each student.

- **Healthcare**: Predictive models in healthcare can anticipate a patient's needs using real-time data, providing proactive care suggestions, alerts, and personalized treatment plans. These systems ensure timely interventions and enhance patient outcomes by addressing potential issues before they escalate.
- **Autonomous vehicles**: Autonomous systems in vehicles analyze complex environments and make proactive decisions to prevent accidents, optimize routes, and enhance passenger safety. These systems operate independently, responding to real-time road conditions and traffic patterns, all while continuously learning to refine their decision-making.
- **Smart homes**: Home automation systems use data from various sensors and inputs to anticipate users' preferences and adjust the environment accordingly. For example, they may adjust temperature, lighting, or other settings before the user explicitly requests them, creating a seamless and efficient living experience.

3.7.2.1 Use Case: Squirrel AI Learning

Squirrel AI Learning is an innovative platform that exemplifies the potential of autonomous and anticipatory systems in the education sector. This system utilizes advanced AI algorithms to deliver personalized learning experiences tailored to each student. By continuously assessing each student's learning patterns, strengths, and weaknesses, the system adjusts the curriculum autonomously to suit their unique needs.

What sets Squirrel AI apart is its ability to predict and address learning gaps before they become obstacles. As the system monitors the student's progress, it proactively adjusts the learning path, ensuring that no student falls behind or goes unchallenged. This anticipatory approach to education enhances learning efficiency, fostering a deeper understanding and retention of knowledge.

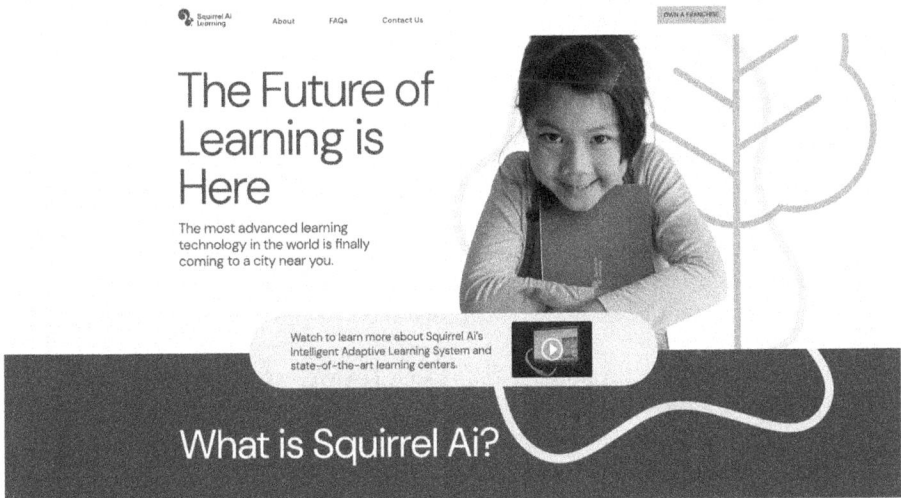

FIGURE 3.8   A screenshot of the Squirrel AI service from 2025.

Source: Available at squirrelai.com

By providing a tailored educational experience at scale, Squirrel AI learning fosters personalized learning environments that enable students to learn at their own pace while receiving appropriate support. This approach has the potential to revolutionize the traditional education model by offering an intelligent, responsive solution that adapts to the evolving needs of each learner.

3.7.3 The Double-Edged Sword: Benefits and Risks

While autonomous and anticipatory systems hold great promise, they also have inherent risks that require careful design and oversight.

Benefits:

- **Proactive problem-solving**: These systems not only respond to user needs but also anticipate them, addressing issues before they occur. This results in improved efficiency, reduced delays, and an enhanced user experience.
- **Scalability**: Unlike processes driven by humans, AI-driven autonomous systems can scale effortlessly, managing thousands or even millions of interactions simultaneously.
- **Increased efficiency and productivity**: By automating complex decision-making processes and repetitive tasks, these systems reduce human workload and enhance operations, ranging from industrial automation to personalized education.
- **Personalized user experiences**: These systems deliver highly tailored interactions by analyzing data in real time.
- **Resilience and adaptability**: Unlike rigid automation, anticipatory AI adapts to new conditions, learns from experience, and continuously improves its decision-making.

Challenges and risks:

- **Ethical dilemmas**: Autonomous systems make decisions without human input, raising significant moral concerns, particularly in high-stakes environments such as healthcare, finance, and autonomous vehicle operations. Who is responsible when an autonomous system makes a harmful decision?
- **Bias and fairness**: If an autonomous system learns from biased data, it may perpetuate those biases in its decision-making process, leading to unfair outcomes.
- **Overdependence on automation**: There is a risk of becoming excessively reliant on these systems, which could result in a decline in critical thinking and human oversight.
- **Privacy concerns**: Autonomous systems that anticipate user needs are frequently built on extensive personal data. As these systems become more familiar with individuals, the risks to data security and privacy also increase.

3.7.4 The Filter Bubble Problem

One significant challenge associated with anticipatory systems—those that make autonomous decisions—is the **filter bubble** [13]. These systems are designed to predict user needs and preferences based on behavioral data. However, in doing so, they can inadvertently limit what users see, hear, and experience. Over time, this creates a digital environment where users are exposed only to content that confirms their existing beliefs, while alternative or diverse perspectives are filtered out.

For example, a content recommendation algorithm on a social media platform might forecast your interests and continue feeding you articles, videos, or posts that align with those preferences. At first, this feels convenient and personal. But gradually, it becomes restrictive—you're surrounded by familiar opinions, never challenged by contrasting views. This echo chamber effect can reinforce cognitive biases, hinder critical thinking, and even contribute to social polarization.

Technically, this phenomenon stems from what's known as **personalization bias**. Algorithms used in content recommendation systems and search engines rely heavily on past user behavior, such as clicks, likes, search history, and watch time, to determine what content to prioritize. While this helps tailor experiences, it also creates a feedback loop: the more a user engages with a particular type of content, the more the system offers similar material. As a result, novelty and diversity are quietly filtered out in favor of relevance and predictability.

One of the most notorious examples of this phenomenon was the Cambridge Analytica scandal during the 2016 US presidential election. The company harvested personal data from millions of Facebook users—often without their consent—to create psychological profiles. These profiles were subsequently used to deliver hyper-targeted political ads, reinforcing voters' biases and influencing their behavior through the use of manipulated emotional cues. It revealed not only the power of data-driven personalization but also the ethical void that can arise when systems optimize for persuasion without accountability.

For designers engaged in anticipatory systems, this poses an ethical imperative: how can we personalize without isolating? How can we maintain relevance without sacrificing discovery? And how can we ensure that our systems empower users rather than subtly curating their worldview?

3.7.4.1 Integrity and Privacy in Autonomous and Anticipatory Systems

Anticipatory systems continuously collect, process, and act on user data—sometimes before the user is aware of it. This can gradually erode user control over choices and experiences—a challenge often referred to as the AI Control Problem [14]. While this enhances convenience, it raises concerns regarding user privacy and data integrity. For instance, smart assistants may anticipate user needs but create unwanted surveillance risks, while autonomous healthcare systems may infer sensitive medical conditions without explicit user consent. To uphold integrity and privacy, designers must:

- **Ensure transparency**: Users should understand what data is being collected, how it is used, and what decisions AI makes on their behalf.
- **Empower users**: Enable them to customize, evaluate, or decline anticipatory AI decisions.
- **Limit data retention**: Make sure the system retains only essential information and removes outdated or unnecessary data.
- **Design for secure access**: Use authentication and encryption to safeguard sensitive AI-driven decisions from unauthorized access.

3.7.4.2 Ethics and Bias in AI

Across all AI patterns, ethical concerns arise when automated systems make decisions that affect people's lives—from hiring to law enforcement to healthcare. Bias, unfair outcomes, and a lack of accountability can perpetuate discriminatory AI inequalities. To ethically design AI, teams must:

- **Adopt fairness-aware design**: Ensure datasets represent diverse populations and avoid reinforcing historical biases.
- **Ensure AI decisions are explainable**: Users should be able to understand and question AI-driven outcomes.
- **Incorporate ethical considerations in development**: Work with ethicists, regulators, and a diverse range of stakeholders when designing AI models.
- **Establish opt-out mechanisms**: Permit users to challenge or dispute AI decisions when necessary.

3.7.5 Leverage Explainable AI

Transparency plays a crucial role in establishing trust in AI systems, especially those that forecast and influence user behavior. XAI enables users to understand the reasoning behind an AI system's decisions, thereby alleviating concerns about the opaque aspects of AI technologies. By integrating XAI principles, we ensure that AI systems are not only efficient but also reliable, adhering to the principles of a human-centered design (HCD) approach.

Specific goals include the following:

- Identifying best practices for integrating XAI into anticipatory systems to support user understanding of AI decision-making processes
- Developing guidelines for presenting explanations that are clear, concise, and meaningful to users without overwhelming them with technical complexity
- Evaluating the impact of XAI-driven explanations on user trust, satisfaction, and willingness to engage with anticipatory systems

3.8 SHAPING THE FUTURE OF AUTONOMOUS AND ANTICIPATORY SYSTEMS

Autonomous and anticipatory systems represent the future of AI, offering profound possibilities for innovation and transformation. Whether it's personalized education through systems like Squirrel AI or self-driving cars navigating roads, these systems are already pushing boundaries and setting new standards for what's possible with AI.

However, like any powerful technology, implementing these systems requires careful consideration of ethical implications, transparency, and fairness. Designers and AI practitioners play a crucial role in ensuring that autonomous and anticipatory systems are developed responsibly, creating solutions that are not only effective but also aligned with human values and societal needs.

In the upcoming chapters, we will explore the architecture, design patterns, and applications of autonomous and anticipatory systems. We will examine how these systems are reshaping industries, influencing our interactions with technology, and the necessity of designing with a commitment to fairness, ethics, and inclusivity.

TL;DR

AI is more than a buzzword—it's a collection of patterns that power intelligent, responsive, and personalized systems. This chapter outlines **seven core patterns** that form the backbone of modern AI applications. Understanding these patterns enables designers and strategists to apply AI thoughtfully, ethically, and effectively across various industries.

Hyperpersonalization—It enables AI to deliver tailored content, recommendations, and interactions by analyzing user behavior and preferences. From media platforms that curate content feeds to e-commerce systems that predict purchase intent, this AI approach enhances engagement and user satisfaction.

Key design considerations:

- Ensure transparency in personalization mechanisms to build trust.
- Provide users with **control** over recommendations and data usage.
- Avoid the **"Cold-Start" problem** by designing fallback experiences for new users with limited data.

Recognition—Recognition-based AI systems identify patterns in images, speech, and text, enabling applications such as facial recognition, medical diagnostics, and voice assistants. These systems significantly enhance efficiency and automation, but they also raise concerns about privacy and ethics.

Key design considerations:

- Design opt-in and opt-out mechanisms for the use of biometric data.
- Address accuracy biases to prevent misclassification, especially in diverse user groups.
- Implement feedback loops that enable users to correct errors and refine the AI's learning.

Patterns and anomalies—AI excels at identifying trends, deviations, and outliers, making it crucial for fraud detection, cybersecurity, and predictive maintenance. These systems continuously learn from historical data to differentiate between normal and abnormal behaviors.

Key design considerations:

- Clearly communicate when and why AI flags an anomaly.
- Balance false positives and false negatives to minimize unnecessary alerts.
- Provide explainability tools so users understand anomaly detection logic.

Conversation and human interaction—This pattern enables machines to engage in natural, human-like interactions, ranging from chatbots and virtual assistants to AI-driven educational tools. However, if AI-generated responses come across as unnatural, it can lead to discomfort, known as the *Uncanny Valley effect*.

Key design considerations:

- Ensure that **AI responses align with user expectations**, such as formal versus casual tones.
- Use **progressive disclosure** to manage user expectations regarding AI capabilities.
- Avoid excessive anthropomorphism to **reduce the Uncanny Valley effect**.

Predictive analytics (PA) and decisions—PA enables AI to analyze historical data and trends, allowing it to forecast future events and assist businesses with risk assessment, financial forecasting, and demand prediction. While powerful, these systems can inherit and reinforce biased data.

Key design considerations:

- Regularly audit datasets to detect and correct bias.
- Make sure users understand that predictions are probabilistic rather than certainty.
- Design interventions for when predictions lead to unintended consequences.

Goal-driven systems—Unlike reactive AI models, goal-driven systems proactively work toward predefined targets, such as autonomous navigation, strategic decision-making, or personalized learning paths. However, these systems run the risk of suffering from the "Halo Effect," where users assume that AI is equally competent in all areas.

Key design considerations:

- **Clearly define AI's scope** so users don't overtrust its decisions.
- Incorporate **explainability features** to demonstrate how AI weighs various factors.
- Ensure that AI can **dynamically adapt to changing contexts** and goals.

Autonomous and anticipatory systems—The most advanced AI patterns, these systems operate with minimal human intervention, predicting user needs and acting before explicit input is required. While incredibly powerful, they also raise concerns related to privacy, control, and the "Filter Bubble" effect, where users may become trapped in AI-curated experiences.

Key design considerations:

- Provide **transparency and empower users** in AI-driven decisions.
- Avoid excessive filtering of content to **preserve diverse perspectives**.
- Prioritize **privacy protection** and **ethical safeguards** in autonomous decision-making.

ML for Designers

4

Learning Phases, Key Concepts, and Models

This chapter aims to equip designers with essential ML principles, enabling them to collaborate effectively with data scientists and engineers to create meaningful anticipatory systems. By gaining foundational knowledge of ML techniques, algorithms, and model training processes, designers will be better positioned to ensure that human-centered design values are integrated effectively into complex AI solutions, particularly anticipatory systems.

Anticipatory systems represent some of the most sophisticated AI applications today. These systems continuously predict and proactively adapt to user needs, relying heavily on advanced data science techniques and rigorous engineering processes. Consequently, designers involved in creating anticipatory experiences must familiarize themselves with core data science and engineering concepts. This foundational understanding fosters clearer communication, productive collaboration, and informed decision-making when working alongside engineers and data scientists.

In previous chapters, we introduced fundamental AI and ML concepts. However, integrating advanced AI, particularly anticipatory systems, into design often creates tension with traditional HCD methods. At its core, HCD prioritizes empathy, user context, transparency, and iterative feedback. Designers focus on qualitative insights, emotional responses, and direct user interactions to deeply understand immediate user needs.

Consider preparing a family dinner. A data scientist might analyze available resources such as ingredients and budget constraints, optimizing selections based on data. Meanwhile, a designer would prioritize the family's tastes, dietary needs, and overall dining experience. Effective anticipatory design requires harmonizing both perspectives—leveraging data-driven insights alongside an empathetic understanding of human desires. Bridging this gap ensures that AI-driven solutions genuinely enrich user experiences rather than merely providing technically "optimal" yet contextually irrelevant outcomes.

DOI: 10.1201/9781003642800-5

In contrast, AI development—especially anticipatory systems—depends significantly on extensive, high-quality datasets, quantitative analysis, and algorithmic accuracy. AI typically demands substantial upfront efforts in data collection, model training, and validation, often before meaningful user feedback can be gathered. This can be counterintuitive to designers used to rapid prototyping, direct user testing, and iterative refinement.

Designers must learn how to bridge empathy-driven methodologies with data-driven engineering practices, advocating for human-centric solutions while respecting technical realities. Achieving this balance requires adopting hybrid design processes, establishing a shared vocabulary, and fostering collaborative working models that address both human values and AI requirements.

Therefore, this chapter will establish a foundational understanding of ML, which is essential for designers seeking to collaborate effectively on anticipatory projects. We will explore the three core phases: data design, model design, and output design, along with the critical steps involved in each phase.

4.1 THREE ESSENTIAL PHASES OF ML

Any ML project typically follows three essential phases:

- **Data design**: Gathering and preparing data.
- **Model design**: Building, training, and evaluating the model.
- **Output design**: Defining how the model's results are presented and iterated upon.

Each phase is essential and interconnected, directly influencing the performance of the ML model and, ultimately, the quality of user experiences.

Within these phases, designers and engineers collaboratively progress through a consistent set of **six essential steps: data collection, data cleaning, model building, model evaluation, generating model outputs**, and **model iteration** [15]. Understanding this structured flow is crucial for designers, as each step represents an opportunity to align user-centered principles with technical requirements, ensuring that ML or anticipatory solutions are both effective and meaningful.

Let's explore each phase in depth, focusing on its impact on both outcomes and the user experiences you design.

4.2 PHASE 1: DATA DESIGN

Think of data as the raw ingredients in a recipe—without quality ingredients, even the best chef cannot create an exceptional dish. Likewise, high-quality data is

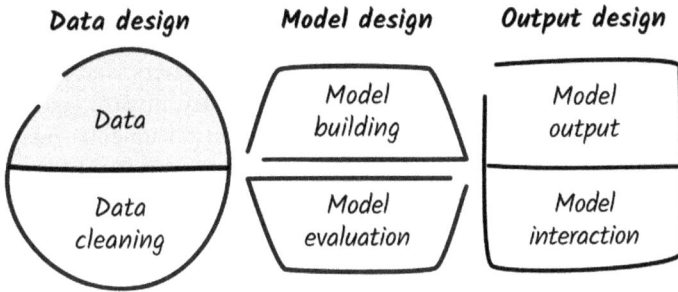

Data design   **Model design**   **Output design**

Data | Model building | Model output
Data cleaning | Model evaluation | Model interaction

FIGURE 4.1   The building blocks of a machine learning workflow focus on the data step.

essential for ML to deliver valuable and accurate insights. But have you ever considered exactly where your data comes from and how its organization impacts your user experience?

Data quality and clarity not only determine the success of your ML models but also influence the quality perceived by users. As designers, a deep understanding of various data types and the necessary preparation can significantly impact how you create intuitive and effective user experiences.

4.2.1 Types of Data

Every piece of data tells a story, but how clearly that story unfolds depends significantly on the type of data you're working with. Let's explore the main types and see how each shapes the narrative of your AI solutions.

4.2.1.1 Structured Data

Structured data is well organized, similar to books on a neatly arranged shelf, making it easy for humans and algorithms to access and interpret. It conforms to predefined models, such as databases and spreadsheets (e.g., Excel or SQL files). Common examples include customer databases containing names, dates, transaction amounts, and patient records listing age, weight, and medical history.

- **Key characteristics:**
  - Clear and predefined format
  - Easy to search and analyze
  - Focused on *what* is happening
- **Stored in** relational databases and data warehouses
- **Organization format** in rows and columns
- **Examples** include dates, phone numbers, customer names, and transaction details.

4.2.1.2 Unstructured Data

Imagine entering an attic filled with old photographs, letters, and a variety of miscellaneous objects. Unstructured data is similar—messy, diverse, and lacking clear organization or purpose. This category includes text documents, images, emails, social media posts, video files, and other data that lacks a fixed format. While it has traditionally been challenging to analyze, modern ML algorithms enable us to quickly and efficiently extract meaningful insights from these previously underutilized resources.

- **Key characteristics:**
  - No predefined model or clear structure
  - Challenging to search and analyze manually
  - Provides context about *why* something happens
- **Stored in** applications, data warehouses, and data lakes
- **Organization format** in various formats, often inconsistent (PDFs, videos, images)
- **Examples** include emails, conversation transcripts, customer feedback forms, and photographs.

The image illustrates the key difference between structured and unstructured data. However, what's particularly interesting is that Gartner estimates only about 20% of corporate data is structured. Although comprising only a fifth of all data globally, this structured portion forms the backbone of what is widely recognized as "Big Data."

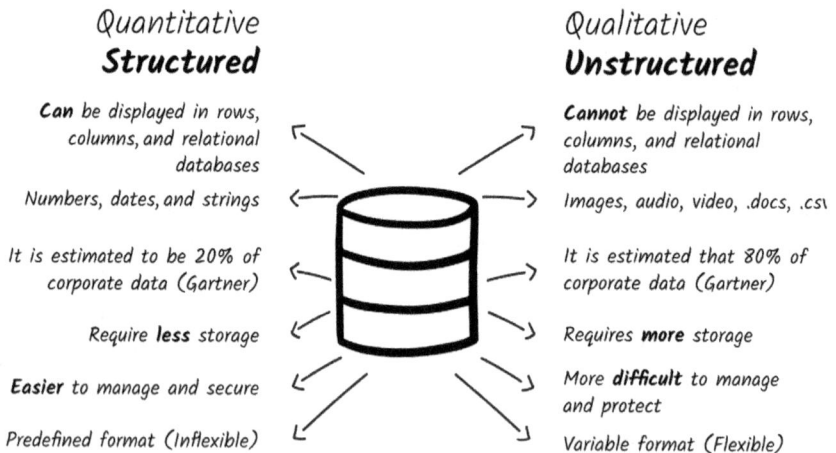

Quantitative **Structured**		*Qualitative* **Unstructured**
Can be displayed in rows, columns, and relational databases		*Cannot* be displayed in rows, columns, and relational databases
Numbers, dates, and strings		Images, audio, video, .docs, .csv
It is estimated to be 20% of corporate data (Gartner)		It is estimated that 80% of corporate data (Gartner)
Require **less** storage		Requires **more** storage
Easier to manage and secure		More **difficult** to manage and protect
Predefined format (Inflexible)		Variable format (Flexible)

FIGURE 4.2   Key features between structured and unstructured data.

4.2.1.3 Semi-Structured Data

Semi-structured data exists comfortably between these extremes. It does not strictly adhere to a fixed structure like structured data, but it is also not completely unorganized. Instead, it utilizes a flexible organizational scheme, such as XML or JSON, enabling adaptability while preserving some structure.

- **Key characteristics:**
  - Flexible yet organized through tags and labels
  - Blends structure with flexibility to accommodate different contexts
- **Stored in** relational databases and tagged-text formats
- **Organization format** in tagged data (HTML, XML, JSON)
- **Examples** include emails sorted into folders (inbox, sent, drafts), server logs, and hashtags.

4.2.2 What Does Data Cleansing Involve?

Before an ML model can effectively learn, the data requires meticulous preparation—a process known as **data cleaning**. This entails carefully refining raw data and transforming it into clear, precise, and valuable information. Think of it as tidying your kitchen before cooking: you remove clutter, clean your tools, and carefully prepare your ingredients.

- **Fixing errors**: Removing typos, correcting inaccuracies, and eliminating unnecessary characters.

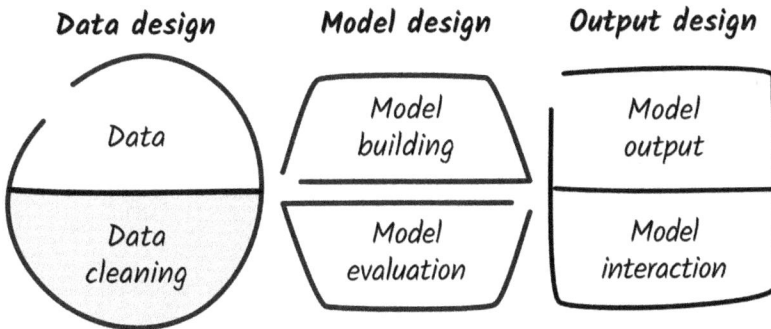

FIGURE 4.3 The building blocks of a machine learning workflow focus on data cleaning.

- **Removing irrelevant data**: Excluding data not relevant to the model's objectives.
- **Handling missing values**: Deciding whether to replace, remove, or adjust missing data points.
- **Managing outliers**: Addressing unusually high or low values that could skew results.
- **Normalization**: Ensuring data consistency by placing it onto a standard scale (like converting various currencies into one).

4.2.2.1 Normalizing Data: Why It Matters!

Normalization transforms data into a format that the ML model can interpret reasonably, preventing extreme values from disproportionately affecting the results.

- **Structured data**: Normalization refers to the process of scaling numeric values to prevent the model from being biased by unusually large or small numbers.
- **Unstructured data**: Normalization involves transforming messy, raw data, such as text or images, into a clear format that ML models can easily interpret and analyze. For instance, free-form user reviews can be categorized by sentiment (positive, negative, or neutral), or images can be converted into numerical pixel patterns that the ML model can process efficiently.

Different data types require specific normalization methods:

- **Text (String)**: Words or categories (such as product descriptions)
- **Whole numbers (integers)**: Counts and quantities (e.g., number of visits)
- **Decimal numbers (float)**: Continuous values (e.g., pricing, ratings)

Without normalization, models can yield misleading results, unfairly highlighting larger numerical values and reducing prediction accuracy. Effective normalization ensures fairness, accuracy, and reliability, fostering trustworthy experiences.

Once cleaned and normalized, your data is ready for ML training. To do this, the data is divided into two sets:

- **Training set (70%)**: Used to teach the model patterns.
- **Testing set (30%)**: Used to evaluate how well the model predicts unseen data.

This split helps ensure that the model generalizes well to real-world scenarios, providing accurate predictions beyond the initial dataset.

4.2.2.2 Why Does Data Design Matter to Designers?

Understanding data preparation is crucial for designers. Poor data quality directly harms user experiences, leading to irrelevant recommendations, inaccurate predictions, frustrating interactions, or even potential harm, undermining the interactions you carefully craft. By actively engaging with data scientists during the data design phase, designers can shape ML systems that are not only effective but also genuinely human-centered.

4.3 PHASE 2: MODEL DESIGN

4.3.1 How Do You Train a Model?

Training an ML model might not be as thrilling as training a mythical creature, but it certainly captures the same sense of adventure and discovery. At its core, training combines rich data and powerful algorithms to create intelligent models that can identify patterns and make informed predictions. How these models learn depends mainly on the "educational approach" we choose. How we teach our models influences how effectively they learn, like guiding a student or training a dragon. Three main approaches shape this learning journey: **supervised**, **unsupervised**, and **reinforcement learning**. Let's explore how each approach helps models "grow" smarter.

4.3.2 Types of Machine Learning

Algorithms learn much like children: through exploring their environment, experimenting, and adjusting based on feedback. Depending on our chosen teaching method, this process can take various forms. Let's examine each approach briefly:

- **Unsupervised learning**: The model independently explores unlabeled data, uncovering hidden patterns and relationships. Imagine giving someone puzzle pieces without a reference image and asking them to make sense of it. It's

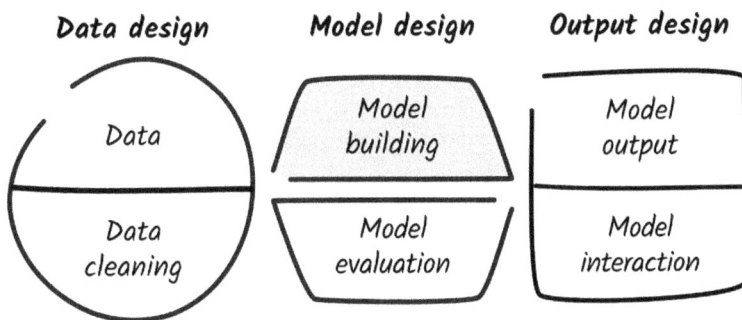

FIGURE 4.4 The building blocks of a machine learning workflow focus on the model-building step.

commonly used in user segmentation, anomaly detection, and market trend analysis.

- **Supervised learning**: The model learns from labeled examples, much like a student practicing with answers at the back of a textbook. It connects inputs with correct outputs, excelling in tasks such as image classification, text recognition, and financial forecasting.
- **Semi-supervised learning**: This can be viewed as a cost-effective version of supervised learning, leveraging a small amount of labeled data alongside extensive unlabeled data. This approach is particularly beneficial when manual data labeling is expensive or impractical.
- **Reinforcement learning**: This method simulates learning through trial and error, using a system of rewards and penalties. Imagine training a dragon— rewarding positive actions and addressing undesirable behaviors. This learning style is widely used in fields such as robotics, gaming, and various process optimizations.

Each method has its strengths and can often be combined strategically to enhance a model's capabilities.

4.3.2.1 Unsupervised Learning

Imagine providing your algorithm with a puzzle without a picture on the box. That's essentially what unsupervised learning is—independently searching for patterns, structures, and groupings within the data. It's crucial for tasks such as segmenting users into meaningful groups, detecting unusual activity, or uncovering hidden market trends. Unsupervised learning typically involves two main types of data mining problems.

Clustering: Imagine crafting a mood board where you intuitively group images by visual similarities, moods, or themes, without predefined categories. The algorithm operates similarly, independently identifying groups within your data. This method is particularly beneficial for tasks such as segmenting customers based on their shopping habits or organizing visuals by dominant colors, styles, or aesthetic preferences.

Association rules: This technique identifies relationships between different data elements. For instance, if a customer buys bread, what's the likelihood that they'll also purchase butter? It's similar to uncovering "if . . . then . . ." patterns, such as, "If a user clicks this button, they're likely to click another one as well." It aids in understanding which elements frequently appear together or influence one another.

Unsupervised learning commonly employs several key algorithms, primarily including:

Neural networks: As discussed in Chapter 1, neural networks are versatile tools that can autonomously discover complex patterns. In unsupervised learning, they can identify subtle visual similarities within datasets, such as grouping images by style or content, without explicit guidance.

K-means clustering: As mentioned earlier, clustering algorithms intuitively group similar data points. K-means effectively divides data into clear and

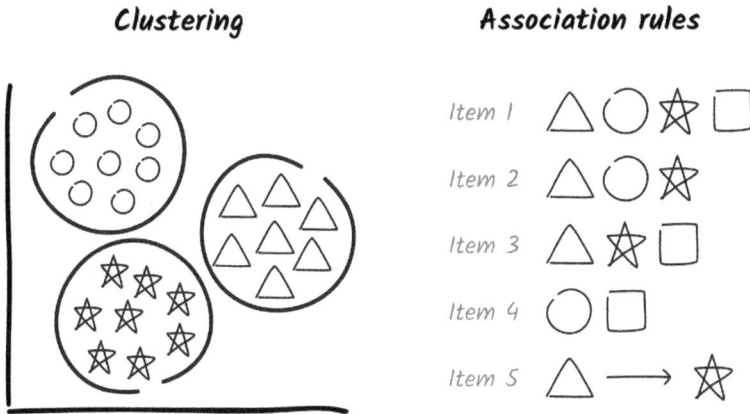

FIGURE 4.5   Visual representation of clustering and association rules concepts.

distinct groups according to their similarities, making it particularly useful for market segmentation, such as grouping customers based on their purchasing behaviors and interests.

Hierarchical clustering: Consider this as organizing your data into a family tree, creating groups and subgroups to uncover relationships and hierarchies. This method helps designers and researchers explore data structures across multiple levels, from broad categories to specific subgroups.

Anomaly detection: Similar to identifying a single mismatched color or shape on a mood board, anomaly detection recognizes data points or patterns that significantly deviate from the norm. These algorithms are crucial for uncovering fraudulent transactions or quality control issues before they escalate.

Principal component analysis (PCA): PCA simplifies complex datasets by highlighting the most influential features, much like selecting key images or themes for a design board. This helps maintain clarity and focus, reducing data complexity without sacrificing essential insights.

What's the designer's role in unsupervised learning? Essentially, it involves three key responsibilities:

- **Aligning user experience with cluster granularity**: Collaborate closely with data scientists and engineers to ensure that the chosen granularity level (how broad or detailed the user segmentation is?) effectively supports a meaningful user experience.
- **Interpreting and presenting patterns**: Create intuitive visualizations or interactions that help users easily understand the insights discovered by the model.
- **Testing for hidden biases**: Actively working with technical teams to identify and mitigate unintended biases in emerging patterns and segments.

By thoughtfully engaging in these tasks, designers ensure that ML models provide all users with relevant, fair, and intuitive experiences.

4.3.2.2 Supervised Learning

Reflect on learning something new with the assistance of a teacher—someone who guides you with examples, corrects your mistakes, and helps you connect the dots. That's the essence of **supervised learning**. This approach trains a model on **labeled data**, where each input is paired with a correct output. Its goal? To learn those associations sufficiently well to make accurate predictions when confronted with new data.

To understand the limitations of supervised learning, consider the story of *Jack*, the little boy in the 2015 film *Room*. Raised in total isolation, Jack's entire understanding of the world is based on what he sees in the small room where he lives—and what his mother tells him. For him, *Room* is the entire universe. Similarly, a supervised learning model can only learn from the examples it is given. If the data is narrow, biased, or incomplete, the model's view of the world becomes equally limited. It may learn to make predictions, but those predictions might represent only a small, skewed version of reality. Supervised learning is typically categorized into two data mining problems.

Classification: This involves sorting data into categories. For example, a model may classify house prices as *below average*, *average*, or *above average*, depending on the location and size of the house.

Regression: This involves predicting a continuous value. Using the same house example, the model could estimate the *precise market price* of a home based on its features.

Both classification and regression involve fitting data to a model that either separates data into categories or illustrates a trend. The choice between the two depends on the type of prediction needed: categorical labels for classification or numerical values for regression. Within supervised learning, several key algorithms are frequently used. The primary ones include the following:

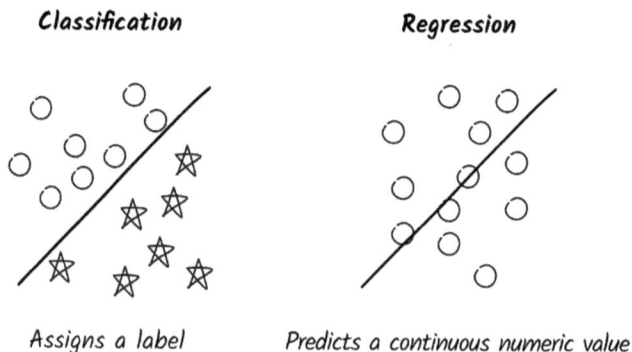

Classification	Regression
Assigns a label	Predicts a continuous numeric value

FIGURE 4.6   Visual representation of classification and regression concepts.

Linear regression: Envision drawing a straight line through your data to predict continuous values, such as forecasting house prices based on square meters or square footage. It's straightforward, transparent, and an excellent starting point for numerical predictions.

Logistic regression: Despite its name, this algorithm does not focus on predicting numbers but rather on categorizing outcomes. It is used for binary classification tasks, assisting the model in making yes/no or true/false decisions, such as determining whether a transaction is fraudulent or whether an email is spam. It is particularly useful when there are only two possible outcome categories.

Decision trees: These resemble flowcharts that guide decisions step by step. They are highly interpretable and suitable for both classification (e.g., determining whether a loan should be approved) and regression tasks.

Neural networks: In Chapter 1, we explore neural networks in detail. Inspired by the human brain, these models excel at identifying complex, nonlinear relationships in large datasets. From recognizing faces to predicting stock trends, neural networks provide exceptional power, though at the expense of transparency. They are robust algorithms applicable in both supervised and unsupervised contexts.

4.3.3 In Short: Designer's Role Across Learning Approaches

Designers may already be familiar with decision trees, but gaining a deeper understanding of regression and classification is essential for shaping truly effective predictive experiences. In the context of supervised learning, designers have three critical roles to play:

- **Designing intuitive data inputs**: Craft interfaces that make it easy and natural for users to provide labeled data, a crucial component of training accurate models.
- **Making predictions understandable**: Present model outputs transparently and trustworthily, helping users make sense of the predictions rather than feel confused by an opaque "black box."
- **Safeguarding data quality**: Collaborate with data scientists to review and refine datasets, ensuring that the model learns from accurate, inclusive, and bias-aware examples.

When designers meaningfully participate in the supervised learning process, the outcome is more reliable models and AI experiences that are more ethical, accessible, and aligned with genuine human needs. Whether ML models depend on supervised or unsupervised data-driven discovery, the designer's role remains crucial. It's not just about making AI functional—it's about making it comprehensible, trustworthy, and human-centered. Designers are uniquely positioned to:

- **Explaining personalization**: Help users understand how and why recommendations are tailored to them.
- **Empowering user control**: Offer clear ways for users to adjust, refine, or dismiss AI-driven suggestions.
- **Ensuring clarity**: Present insights and outputs in a way that builds trust and avoids confusion or mistrust.
- **Balance accuracy with transparency**: Collaborate with technical teams to strike a balance between performance and user experience.
- **Design for control**: Build interfaces that let users oversee, guide, or intervene in automated decisions.
- **Simplify complexity**: Translate the invisible logic of ML into interactions that feel natural, transparent, and empowering.

This holistic approach ensures that ML remains a powerful tool to enhance user experiences, not a source of friction or uncertainty.

4.3.3.1 Semi-Supervised Learning

Semi-supervised learning strikes a balance between supervised and unsupervised methods by utilizing limited labeled data in conjunction with abundant unlabeled data. Imagine a teacher giving initial guidance but then stepping back, allowing the student to explore and learn independently. This approach is particularly beneficial when labeling large datasets is costly or impractical. Consequently, the designer's role in semi-supervised learning includes three key responsibilities:

- **Ensuring data diversity**: Help validate if the labeled dataset represents the diversity of actual users, ensuring inclusive and equitable learning outcomes.
- **Validation interfaces**: Design user-friendly interfaces that allow users to review and correct incorrect predictions, thus continuously refining the model.
- **Facilitating data annotation**: Develop simple, intuitive interactions that encourage users to easily provide additional labels, improving the model's accuracy and relevance.

4.3.3.2 Reinforcement Learning

Reinforcement learning (RL) is the most dynamic form of ML. Instead of learning from static datasets or uncovering patterns in raw information, **reinforcement learning trains an** *agent*—a decision-making system—to interact with its environment and learn from the consequences of its actions. Every decision the agent makes is followed by feedback: a *reward* if the action is beneficial or a *penalty* if it isn't. Over time, the agent optimizes its behavior to maximize rewards and minimize errors.

This is not just about pattern recognition—it's about building systems that learn *through experience*, adapting their behavior as they go. That's why RL is widely used in areas where real-time adaptation is essential, such as robotics, autonomous vehicles, recommendation systems, and complex, adaptive interfaces. Therefore, reinforcement learning is also one of **the most critical building blocks for anticipatory systems**.

As we saw in the previous chapter, while other AI patterns respond to inputs, anticipatory systems act before the user even makes a request. They must comprehend not only the present but also what might come next. Reinforcement learning enables this forward-looking intelligence. It doesn't just teach systems to react better; it trains them to make smarter, more proactive decisions over time. That's the essence of anticipation. This makes reinforcement learning unique because it emphasizes long-term outcomes rather than just immediate feedback.

This transforms our role as designers into a unique and powerful space. We are no longer creating fixed experiences—we are co-creating behaviors. Systems trained through reinforcement learning are continually evolving, meaning the design isn't merely a snapshot in time—it's a living dialogue between the model, the user, and the interface. This is an exhilarating shift but requires more intentional, ethical, and collaborative design practices. Hence, the designer's role in reinforcement learning is crucial and nuanced in many dimensions.

Shaping reward systems thoughtfully: As a designer, you may not be the one coding the reward function, but you should *definitely* be involved in defining what success looks like for users. In reinforcement learning, the system learns to repeat actions that maximize rewards. However, if we only reward superficial metrics, such as clicks or time spent, we risk creating manipulative or addictive systems. Designers must help establish *human-centered* goals, ensuring the system rewards clarity, usefulness, or user satisfaction instead. Your role is to ensure that what's "good" for the algorithm is also genuinely beneficial for the user.

Designing simulated training environments: Reinforcement learning systems are trained in simulations before being deployed in the real world. Designers play a crucial role in ensuring that simulations accurately reflect real-world diversity, complexity, and nuance, rather than just idealized edge cases. That way, the system learns in environments that mirror real user behaviors and needs.

Preventing harmful feedback loops: Since reinforcement learning systems adapt in real time, they are susceptible to reinforcing undesirable behaviors. A system might learn to exploit biases, chase harmful attention metrics, or over-optimize for narrow goals. Designers play a crucial role in auditing, questioning, and shaping the model's evolving behavior, ensuring that it supports users rather than exploits them.

Now that we've seen how crucial it is to define the appropriate reward system, there's another dynamic at play in reinforcement learning: how frequently the system should attempt something new. This introduces us to the *exploration versus exploitation dilemma*—one of the most intriguing challenges in designing anticipatory systems.

4.3.3.3 The Exploration–Exploitation Dilemma

Imagine dining at a new restaurant with an extensive menu: should you explore by trying a new dish or stick to what you know by ordering your favorite meal? Reinforcement learning faces a similar dilemma. Striking the right balance between exploring new strategies and exploiting known successes is crucial for efficient learning. Excessive exploration can lead to missed immediate rewards, while insufficient exploration can hinder the discovery of more effective long-term strategies. Achieving reliable, ethical, and user-aligned models across these learning methods demands thoughtful designer involvement. But how do we strike this delicate balance?

4.3.4 How Do You Design Reward Functions?

The reward function is everything when working with AI systems, especially anticipatory ones. It's what teaches the model what "success" looks like. Yet, in many teams, this fundamental part of development often occurs behind closed doors, often without design input, which is a significant issue.

We must take a brief (but essential) detour through a few statistical concepts to understand how rewards are defined. Don't worry—I'll keep the math light and the relevance high.

4.3.5 First Things First: Understanding Model Errors

Imagine you are developing an AI-powered tool for cancer diagnosis. This system, known as a binary classifier, predicts whether a person has cancer by making one of two predictions: *yes* or *no*.

When this type of model makes a prediction, it results in one of four possible outcomes:

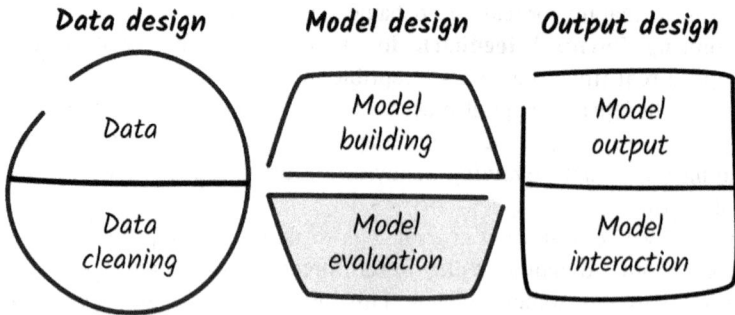

FIGURE 4.7 The building blocks of a machine learning workflow focus on the model-evaluation step.

- **True positive**: The model correctly predicts that a person has cancer when they do.
- **True negative**: The model correctly predicts that a person is healthy when there is no cancer present.
- **False positive**: The model incorrectly predicts cancer in a healthy person.
- **False negative**: The model incorrectly predicts that someone does not have cancer when they do.

Now pause for a moment and reflect. . . . Which scenario do you think is worse for the user: being told they have cancer when they don't, or being told they are fine when they do have cancer?

This represents the **moral and functional dilemma** we encounter when designing intelligent systems. Here, **Type I (false positive)** and **Type II (false negative)** errors become significant.

4.3.6 Precision Versus Recall: What Are We Optimizing For?

This brings us to two important performance metrics: **precision** and **recall**, also known as sensitivity. They assess how effectively a model retrieves **relevant** results and avoids irrelevant ones.

- **Precision** tells us: Of all the people we said to have cancer, how many do?
- **Recall** tells us: Of all the people who have cancer, how many did we catch?

If we **optimize for precision**, we instruct the model to be *very confident before asserting that someone has cancer.* This approach minimizes false positives but risks overlooking actual cases. Conversely, if we **optimize for recall**, we aim to identify *every potential cancer case, even if this results in some healthy individuals being flagged.* This strategy decreases false negatives but increases false positives. Both approaches have trade-offs and necessitate careful consideration from the entire product team, not just data scientists.

4.3.7 Why Should Designers Care?

Let's be clear: the definition of the reward function shapes everything! Whether you are designing for healthcare, finance, mobility, or e-commerce, the behavior of an AI model will directly influence how people experience your product. Yet, too often, designers are excluded from these early decisions, when their input is needed most.

Designing a reward function is not just a technical exercise; it is a **values-driven decision**. It involves determining *what the system should prioritize.* That choice carries ethical, emotional, and practical consequences for those affected by the system's decisions, whether or not they are aware of the AI operating behind the scenes. This is where designers play a crucial role. You provide the human perspective and pose the questions that others may overlook:

- What does a false positive feel like in this context?
- What are the real-world consequences of a false negative?
- How do we communicate uncertainty without causing panic or false reassurance?

Designers transform data-driven decisions into meaningful human outcomes. This means we cannot afford to remain silent at the table where those decisions are made.

4.3.8 Case Study: Designing Reward Functions in Google's Read Along App

A powerful example of how reward functions shape user experience comes from Google's Read Along app (formerly known as Bolo). This mobile application helps children learn to read using only their voice. In Episode 9 of Google's *Centered* YouTube series [17], the team revealed that optimizing solely for pronunciation accuracy—a high-precision metric—led to unintended consequences. When the system failed to recognize a child's correctly spoken word (resulting in a false negative), it often left the child feeling frustrated and stuck. This was particularly true for children with a completionist mindset who wanted to "turn all the words green" before moving on to the next task.

To address this issue, the team redesigned the reward function to prioritize progress over perfection. Instead of requiring each word in a sentence to be pronounced flawlessly, the app now rewards partial success, allowing children to continue reading even if they get just a few words right. The result? A transition from rigid correctness to joyful learning. This subtle yet powerful change had a measurable impact on motivation, confidence, and emotional engagement.

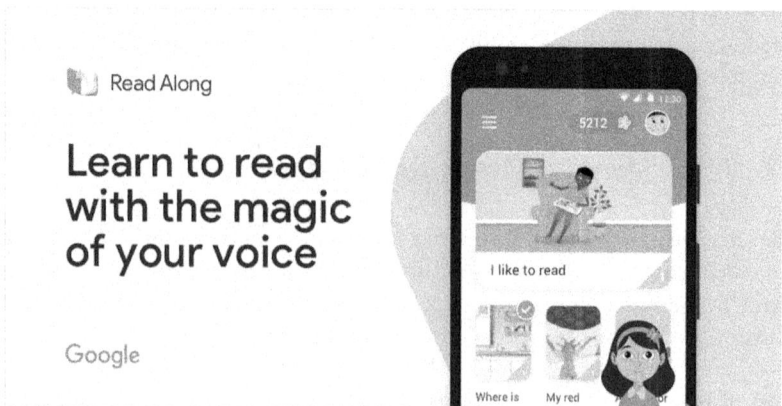

FIGURE 4.8   A 2025 screenshot of Google's Read Along app [16].

This case reminds us that defining a reward function isn't merely a statistical challenge—it's also a design one. Designers play a vital role by asking the right questions:

- What does a false negative *feel* like to a child trying to learn?
- What message does the interface send when it says, "You failed," even when the child didn't?
- How can we strike a balance between technical accuracy and emotional encouragement?

In Read Along's case, human-centered thinking transformed a model fixated on precision into one that fosters curiosity and celebrates growth—an essential reminder for anyone involved in designing AI-powered learning experiences.

4.3.8.1 Beyond Accuracy: The F1-Score and Balanced Evaluation

Someone might say, "Our model is 95% accurate!" That sounds impressive—until you realize that in a dataset of 100 people, 95 are healthy. A model could predict "no cancer" for everyone and still achieve 95% accuracy. However, it would miss every actual case of cancer.

This is why **accuracy alone can be misleading**, particularly in imbalanced datasets where one outcome is significantly more common than the other. This is where the **F1 Score** comes into play. It is a metric that balances **precision and recall**, offering a more comprehensive view of the model's performance, especially in high-stakes situations where the costs of errors are uneven.

Designing a reward function isn't merely a technical step—it's a **collaborative design challenge**. It requires collaboration among product managers, data scientists, engineers, and designers.

When approached thoughtfully, it establishes the foundation for AI that is not only intelligent but also responsible, transparent, and aligned with human needs and motivations. As a designer, your role is to ensure that reward functions optimize not just statistical success but also real-world relevance, user trust, and long-term value.

4.4 PHASE 3: OUTPUT DESIGN

Now that the model has been trained, what emerges from the other end? This is where **output design** comes into play—the final, crucial phase of any ML training, where predictions are surfaced and decisions are made based on them. It is where the model's logic meets human experience. Moreover, it is where designers play a vital role in shaping how machine intelligence becomes visible, usable, and trustworthy.

At the heart of output design is a concept known as the **confidence level**. You've likely seen this in real life—when someone attempts to guess another person's age, they

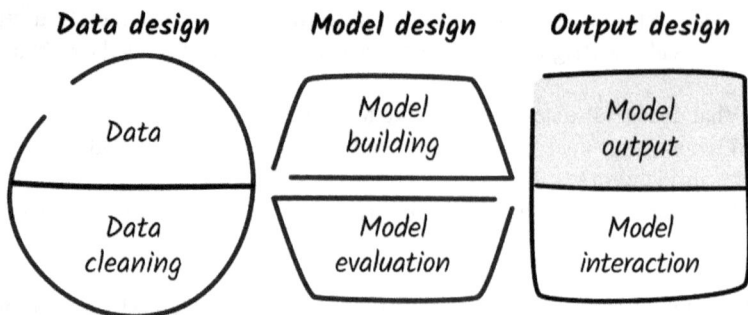

FIGURE 4.9 The building blocks of a machine learning workflow focus on the model-output step.

might say, "I think she's 32. . . . maybe 35?" That hesitation indicates low confidence. ML models operate in much the same way. They don't simply state, "This person is 35." Instead, they assert, "I'm 82.2% confident this person is 35."

That confidence level matters—a lot! It determines when a model acts independently and when it should pause for human review. For example, if an email is 97% likely to be spam, it can be safely moved to the spam folder automatically. However, if the confidence level is only 55%? In that scenario, a careful system might prefer to consult the user first. This balancing act between certainty and uncertainty is fundamental to the development of responsible AI.

4.4.1 Why Confidence Doesn't Equal Certainty?

One of the most important aspects to understand about ML outputs is that there is no such thing as 100% certainty. ML systems are built on statistical patterns, not facts—they function as probabilistic machines trained to recognize trends and make predictions based on likelihood. Even highly optimized models can produce incorrect answers, and sometimes with unshakable confidence. That's why communicating uncertainty is not merely a technical responsibility but rather a design challenge.

Even the most accurate systems, those claiming 97% or 99% precision, still leave room for error. A small percentage of errors can lead to serious harm in high-stakes environments, such as healthcare, finance, mobility, or education. In fact, data extraction accuracy averages around 80% across most real-world solutions, meaning that roughly one in five outputs may be incorrect. These mistakes can be identified and corrected in systems where users are invited to review, adjust, or override outputs. But what occurs when humans are *not* involved?

This challenge becomes even more critical in anticipatory systems, where decisions and actions are taken *without user input or awareness*. In these experiences, users aren't asked to confirm or validate—they simply experience the result. As a consequence, confidence thresholds take on a new level of importance.

If a system makes a confident decision but fails, the user may never understand why it occurred, or worse, may start to distrust the system entirely. Since these systems aim to operate with little resistance, they typically lack a fallback mechanism or user correction loop.

In this context, confidence transcends mere numbers—it becomes a signal of risk, trust, and accountability. Designers must collaborate closely with data scientists to determine when the system has sufficient confidence to act independently, when it should pause or defer, and how to manage uncertainty in a way that safeguards the user experience without overwhelming it.

Ultimately, anticipatory systems increase the responsibility of both developers and the design team to ensure accuracy, given the minimal user involvement. The margin for error becomes much smaller, and the consequences of mistakes—whether a false positive or a false negative—become even more serious. If the system misjudges a user's needs because it is overly confident, it could undermine trust, create friction, or even lead to subtle but harmful consequences.

In anticipatory systems, confidence is not just a statistical metric—it's a design responsibility. Designers must ensure that:

- High-confidence outputs are aligned with high-stakes decisions.
- Medium-confidence outputs are treated with caution or flagged for human validation.
- Low-confidence outputs are handled gracefully, transparently, or even withheld to prevent unnecessary risk.

Equally important, **confidence must be communicated**—either directly through the interface or indirectly through system behavior. A proactive system that appears "confident" must indeed *be* sure, or allow for graceful correction.

As we enter this new era of intelligent, proactive products, designers are no longer merely shaping interfaces; they are also shaping the experiences that users have with these products. **We are influencing behaviors, expectations, and outcomes**. In doing so, we must design for a world where assumptions seem invisible but carry significant consequences.

4.4.2 The Iterative Nature of Model Refinement

Designers often view iteration as a prototyping component, but in ML, **iteration serves as the engine**. Model training isn't a one-time task; it resembles sculpting: we begin with a rough form (a basic model) and refine it over time by eliminating errors and optimizing performance. This occurs through repeated cycles of training and testing. As previously noted, approximately 70% of the data is used for training, and 30% is used for validation. This process helps prevent **overfitting**, where the model becomes overly tailored to past data and struggles in real-world applications.

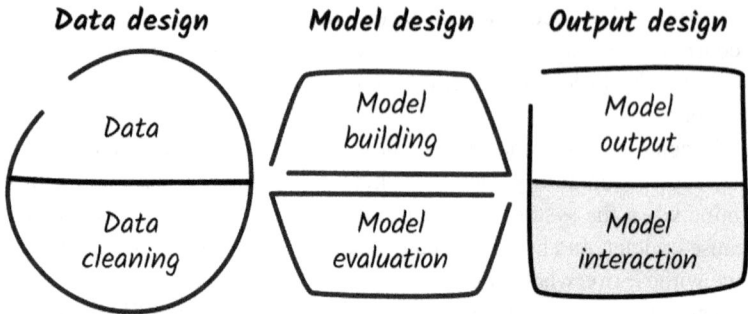

FIGURE 4.10   The building blocks of a machine learning workflow focus on the model-interaction step.

As the model improves, so too should the design. Designers must adapt their approach to presenting confidence, communicating risk, and adjusting system behavior. Continuous feedback loops are essential not only for data scientists but also for experience design.

4.4.3 From Confidence to Collaboration: Human-in-the-Loop

Designers can rely on a robust strategy: human-in-the-loop (HITL) when a model is uncertain or when the stakes are high. HITL incorporates human input, oversight, and judgment within the ML decision-making process. It serves as a way to add guardrails and compassion to the automation process. In practice, HITL can take various forms:

- A radiologist reviewing AI-flagged X-rays before diagnosis.
- A content moderator double-checks flagged posts.
- A system asks a user to verify whether a recommendation is relevant.

Designers play a crucial role here. They determine when to involve humans, how to request input, and how to convey uncertainty without overwhelming or misleading the user. HITL is about trust, not just control—it's about demonstrating to users that they're not being sidelined by automation but supported by it.

4.4.3.1 HITL: Designing for Uncertainty, Together

We now understand that in ML, one uncomfortable truth remains: confidence is not the same as certainty. Even the most sophisticated models are probabilistic by nature, and their predictions can still be incorrect, despite how confident they may seem. That's why human oversight is not merely a safety net but an essential aspect of responsible AI.

HITL typically plays a critical role at two points in the ML workflow—at the beginning and end of the decision-making process. If no ready-to-use model fits the problem domain, but large amounts of raw data are available, humans become essential during the data preparation process. They clean, annotate, and structure datasets, transforming noisy inputs into usable training material. This upfront work provides the model with the best possible foundation for learning, much like giving a child the right books before they begin to read.

Even when models are already trained, HITL can operate post-prediction, particularly in high-stakes environments. A model might automatically sort insurance claims or flag financial transactions. Still, a human may step in to confirm edge cases or validate ambiguous results before the system takes irreversible actions.

4.4.4 Forms of Feedback: Explicit Versus Implicit

Human feedback in these systems generally falls into two categories: explicit and implicit.

Explicit feedback: This is deliberate and intentional. A user corrects a recommendation, marks a prediction as incorrect, or affirms a suggested action. This type of feedback is typically accurate and reliable because it is directly stated, similarly to marking an email as "not spam" or clicking "No, this wasn't helpful."

Implicit feedback: This is subtle and inferred. It's based on behavior rather than conscious input. For instance, the time spent on a page, repeated actions, or interaction patterns can indicate satisfaction or frustration, without a word being spoken. Designers should recognize the power and ambiguity of this type of feedback. However, enabling passive learning also introduces risks: are we accurately interpreting the users' actions?

4.4.5 Anticipatory Systems and the Confidence Dilemma

We discussed that in anticipatory systems, where predictions are made and acted upon *before* the user even makes a request, the stakes of prediction confidence are even higher. Since the system takes proactive steps, any error can directly shape the user experience, often without the user's awareness. In this context, HITL becomes a safeguard for agency and trust. Therefore, designers need to consider and make decisions:

- At what confidence threshold should the system act autonomously?
- When should it ask for confirmation?
- How can uncertainty be communicated without creating anxiety?

The boundary between helpful and harmful is thin in anticipatory experiences, and confidence plays a key role in defining that line.

4.4.6 Why Output Design Matters?

This phase closes the loop. The data have been prepared. Models have been trained. However, everything upstream collapses if the output is not usable, understandable, or trustworthy. Therefore, designers need to learn to:

- Translate confidence scores into meaningful cues.
- Help users understand the difference between strong and weak predictions.
- Build interfaces that allow for correction, feedback, or escalation.
- Anticipate edge cases and ethical risks, not just edge designs.

When executed correctly, output design guarantees that AI not only functions but also effectively serves people! This should be the primary goal of the entire process: ensuring that machine intelligence aligns with human reality through clarity, accountability, and care.

TL;DR

ML isn't just a technical challenge—it's a design opportunity. This chapter demystifies how ML works, offering a practical guide for designers to collaborate meaningfully with data scientists and engineers. By understanding the three core phases of ML—**data design, model design, and output design**—designers can help shape more ethical, intuitive, and user-aligned AI systems, particularly anticipatory ones.

Data design—Great AI starts with great data. Designers must understand the types of data (structured, semi-structured, and unstructured) and how it's cleaned, normalized, and split for training and testing.

Key insight: Poor data lead to poor experiences. Designers can play a pivotal role in ensuring that data reflect real-world nuance and diversity.

Model design—Designers don't need to code algorithms, but they do need to understand how models learn via supervised, unsupervised, semi-supervised, and reinforcement learning.

Key role: Guiding data labeling strategies, structuring meaningful reward systems, pattern visualizations, and bias mitigation.

Output design—The way AI communicates confidence, predictions, and errors directly impacts user trust. Especially in anticipatory systems, designers must define when AI acts alone, when it defers to humans, and how uncertainty is framed.

Key principle: Confidence ≠ certainty. Design for transparency, graceful failure, and human-in-the-loop (HITL) validation.

Designing Reward Functions in Reinforcement Learning: In **reinforcement learning**, AI learns by receiving rewards for actions, so the **reward function** defines what "success" means. If that definition is flawed, the system may learn the wrong behaviors.

This isn't just technical—it's a **design decision, too**. Designers need to help define human-centered goals: should the system reward speed, accuracy, emotional engagement, or long-term learning?

For example, the initial reward function in Google's Read Along app focused on perfect pronunciation. However, when the system failed to recognize a child's spoken word, it caused frustration. Redesigning the reward to emphasize progress over perfection made the experience joyful again.

Designers must know how to shape reward systems to reflect what is truly valuable for users, not just what is easy to measure.

Chapter's biggest takeaway is that ML is not a black box—it's a co-design space. When designers shape data flows, reward logic, and feedback loops, AI becomes not just smarter but also more responsible, human-centered, and aligned with real-world needs.

PART II

Designing Relief

Addressing the Cognitive Burden of Technology Through Anticipatory Systems

The Paradox of Choice in the Digital Era

5

In the first part, we laid the groundwork to understand the mechanics behind intelligent solutions. Now, we turn our attention to the current state of technology and its impact on our professional and personal lives, paving the way to expose how anticipatory design may become a key resource to overcome these challenges.

Imagine it's 7:00 am. Your phone, lying quietly on your bedside table, suddenly lights up, and your alarm rings. You reach over to turn it off, and before you've fully awakened, you're bombarded by notifications: six urgent emails from your manager, 323 unread Slack messages, and a friendly nudge from your fitness tracker reminding you to hydrate and take 10,000 steps. And, of course, your social media apps are buzzing. You swipe away the alerts, but more arrive like a persistent, invisible crowd demanding your attention.

As time goes on, you subconsciously make dozens of tiny decisions:

- *What's the weather like today?*
- *Which emails require my immediate attention? And which ones can wait until tomorrow or next week?*
- *Do I eat breakfast or fit in a quick workout?*

As your day unfolds, your cognitive capacity decreases. By evening, choosing something to watch on Netflix feels like a monumental task. The issue isn't just deciding between chicken or fish for dinner. It's a flood of choices that makes even trivial decisions overwhelming.

FIGURE 5.1 Visualizing notification overload on a mobile device.

DOI: 10.1201/9781003642800-7

This isn't just a personal anecdote—it's a shared experience in our digital age. Technology was meant to simplify our lives, but it often creates a mental tax we never asked for. With devices constantly within reach and an unending stream of content, we often find ourselves overwhelmed by the weight of countless decisions. While technology is undeniably powerful, our brains aren't naturally equipped to process this constant influx of data. Instead of enhancing our lives, it often leaves us fragmented, exhausted, and overwhelmed [18].

5.1 WHEN MORE BECOMES TOO MUCH: INFORMATION OVERLOAD

Consider the endless streaming libraries, overflowing online marketplaces, and relentless notifications. While it may seem like a dream, having too many choices can lead to paralysis, indecision, and mental exhaustion. Psychologist Barry Schwartz famously coined the term "paradox of choice," where having more options doesn't lead to more freedom or satisfaction, but rather to regret and decision fatigue [19].

Information overload isn't just an individual struggle—it's a design problem. Tools meant to simplify life often make it more complex. Think about picking a movie from a streaming platform or comparing dozens of seemingly identical products online. What should be a quick, enjoyable process turns into a frustrating experience.

Netflix is a prime example, where the volume of options overwhelms users, resulting in decision fatigue and cognitive overload. Netflix's mission is to deliver entertainment effortlessly at users' fingertips, yet its design frequently undermines this goal.

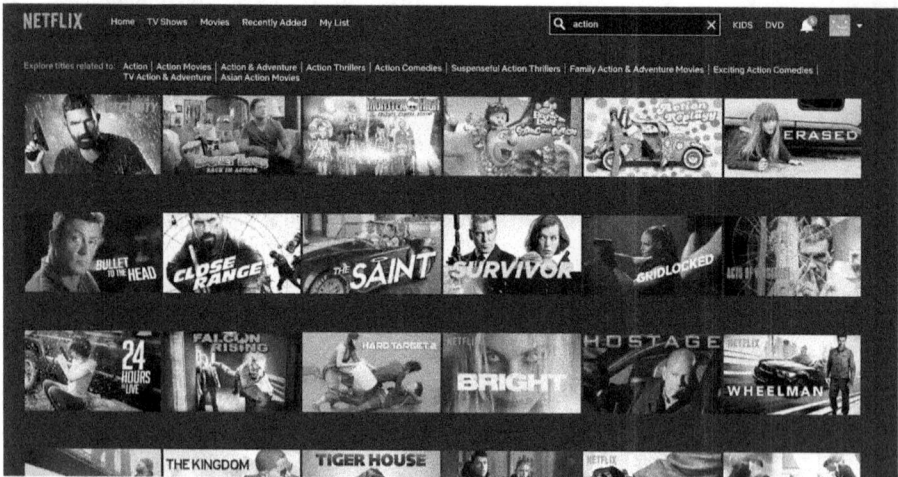

FIGURE 5.2   Representation of a Netflix Homepage view of its portfolio content [21].

Picture a typical scenario: a user logs into Netflix, anticipating a relaxing evening of entertainment. Instead, they find themselves endlessly scrolling through content, unable to make a decision. Studies reveal that Netflix users spend an average of 18 minutes per session browsing rather than watching, turning what should be a seamless experience into a frustrating one [20].

The challenge is not unique to Netflix. Competitors like Amazon Prime take a deliberate approach, prioritizing revenue over minimizing cognitive load. Recently, Amazon introduced an option to rent or buy movies within its streaming platform, adding unnecessary friction and potentially complicating parental controls. While Netflix at least strives to simplify decision-making through algorithms that recommend personalized content, Amazon's strategy intentionally complicates the user journey. By prioritizing business contracts and revenue streams over user satisfaction, Amazon creates a more frustrating and cumbersome customer experience.

This highlights a fundamental flaw: these platforms, along with many other services, often prioritize showcasing content or offers over helping users make informed choices. For users with limited time, the experience becomes counterproductive, creating anxiety and paralysis. Rather than creating an engaging, seamless journey, Netflix's abundance of choices and Amazon Prime's intrusive publicity impose cognitive demands that detract from user satisfaction.

These examples highlight a larger challenge: the information overload in today's digital world increasingly contributes to cognitive overload and decision fatigue, making it more difficult for individuals to focus, take action, or feel emotionally satisfied.

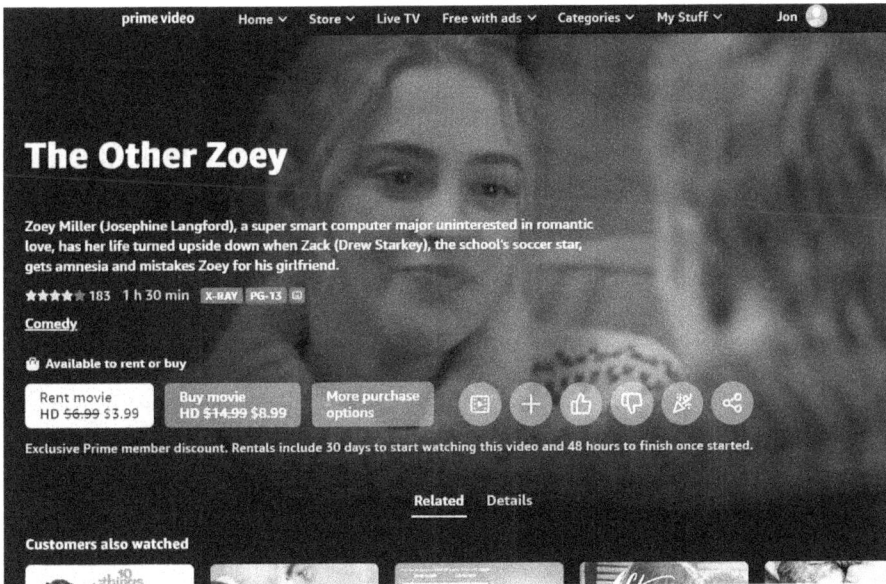

FIGURE 5.3 Amazon Prime features for renting or buying a movie [22].

5.1.1 The Cognitive Costs of Modern Connectivity

Technology has changed how we interact with the world, granting unprecedented access to services, tools, and knowledge. But constant connectivity has introduced a new kind of fatigue. The volume of notifications, options, and to-dos demands our attention at every turn. Schwartz calls this the "tyranny of choice"—a condition that leads not to freedom but to burnout and poor decisions [19].

Cognitive overload occurs when the brain is overwhelmed by excessive information, leading to mental fatigue and a decline in focus [19, 23]. For younger, tech-dependent generations, the effect is even more pronounced [24]. Forbes reported in 2021 that this had been linked to burnout, particularly among highly engaged technology users such as Millennials and Gen Z [25]. Neuroscientist Anne-Laure Le Cunff explains that our brains are often overwhelmed today because the world changes so rapidly. We collect a lot of information in an effort to understand the complex things around us. However, this rush to gather data leads to a misunderstanding: people think that just having information means they truly understand it and can use it effectively [26].

5.1.2 The Illusion of Knowing: Information Versus Knowledge

At its most fundamental level, information comprises isolated facts, superficial details, and disconnected snippets of data. On the other hand, knowledge involves a deeper synthesis—a meaningful integration of information into insights, critical reasoning, and informed decisions. However, in an era defined by instant digital gratification, our brains frequently mistake superficial familiarity for real comprehension. As Anne-Laure describes, this phenomenon is partly driven by our anxiety to keep pace with a rapidly changing world, fueling a false sense of security through constant data collection.

Social media, a powerful driver of exposure to superficial information, significantly contributes to this illusion. Platforms like Instagram, Twitter, TikTok, and YouTube are flooded with influencers and self-proclaimed experts who deliver quick, simplified insights packaged attractively but often lacking depth. Followers, impressed by the immediacy and frequency of content, mistake familiarity and repetition for genuine expertise. The design of social media, which relies on dopamine-driven notifications and engagement algorithms, perpetuates shallow consumption habits and fosters what I refer to as the *Information-Knowledge Illusion Bias*.

I view this as a tendency to mistake superficial exposure to information as equivalent to deep understanding and authentic knowledge. It occurs when the ease of recalling quick facts or frequent exposure to familiar data leads us to believe we've mastered complex topics. This bias leads us away from genuine comprehension toward shallow engagement with information.

I also believe that this issue—*Information-Knowledge Illusion Bias*—uniquely shapes the cognitive development of younger generations, specifically Gen Z and Alpha. These digitally native generations have grown up with continuous internet and social media access, resulting in an unprecedented reliance on instantaneous information. Their constant exposure to simplified, bite-sized content, often delivered by influencers rather than experts, fosters an illusion of comprehensive understanding. As a result, these generations may develop superficial cognitive habits, favoring rapid consumption over deep reflection and critical analysis. This phenomenon poses significant challenges to education and intellectual growth, potentially hindering their ability to engage deeply with complex issues and weakening their resilience to misinformation and cognitive manipulation.

Why do our brains get caught in this trap? A variety of cognitive biases combine to mislead us into confusing information for actual knowledge. The **Dunning–Kruger effect** describes how individuals with limited actual understanding become overconfident, mistaking superficial familiarity for expertise. Another issue is **confirmation bias**, which further exacerbates this problem, as we tend to gravitate toward data that validates existing beliefs rather than engaging critically with complexity. Conversely, we have **information bias**, which underscores our flawed instinct to collect more data under the assumption that more information inherently improves decision-making. Anne-Laure expands this cognitive landscape with the notion of the *maximized brain*, an anxiety-driven mindset in which someone believes every task or goal must be pursued to the extreme. Every piece of information must be absorbed entirely; this mindset amplifies cognitive overload and burnout, creating fertile ground for superficial engagement rather than meaningful learning and discovery [26].

5.1.3 How Anticipatory Design Can Help?

This growing challenge demands innovative strategies that streamline interaction, reduce mental effort, and mitigate cognitive bias. Anticipatory systems, powered by AI, ML, and DL, offer powerful tools to address the effects of the digital era. The idea is simple: reduce unnecessary friction, automate routine decisions, provide guidance at just the right time, and encourage deeper cognitive engagement while minimizing superficial information consumption. When designed thoughtfully, this can ease the mental load without taking away agency.

Aaron Shapiro, an American entrepreneur, suggests that the future of design lies in creating products that anticipate user needs and make decisions on their behalf, delivering guidance precisely when needed without requiring ongoing input [27].

Imagine a to-do list that automatically prioritizes your tasks based on shifting deadlines or a meal-planning app that adjusts your grocery list based on your dietary needs and current pantry inventory. These aren't just intelligent conveniences—they reduce decision fatigue, enhance productivity, and promote healthier habits by aligning everyday choices with your long-term goals. When done right, anticipatory systems:

- **Minimize cognitive overload** by highlighting only the most relevant information.
- **Mitigate decision fatigue** by automating low-stakes choices.
- **Promote better decisions** by framing options around user needs and motivations.

As Shapiro notes, anticipatory systems shine when they simplify, not by adding more features but by getting out of the way.

5.2 THE PSYCHOLOGY AND COST OF CHOICES

Cognitive strain is bad for individual well-being and also for the systems we design. Decision fatigue and analysis paralysis result from cumulative exhaustion caused by making too many choices, undermining the effectiveness of technology designed to help us. Cognitive strain erodes trust, reduces engagement, and diminishes the user experience. In pursuing innovation, designers and technologists have overlooked a critical question: how can technology alleviate, rather than amplify, the mental burden it creates?

When faced with many options, the relationship between decision-making and choice is complex and deeply influenced by cognitive biases, heuristics, and the nature of human behavior. The problem of excessive options is aggravated by cognitive fatigue. Research by Schwartz and others shows that an abundance of options doesn't empower users—it overwhelms them. Faced with too many choices, people often second-guess themselves, experience decision regret, and ultimately feel less satisfied with the outcomes [19]. The cognitive effort required to evaluate numerous options can leave users questioning whether they have made the "best" choice, thereby increasing self-doubt and dissatisfaction. This phenomenon, known as the paradox of choice, underscores the need for systems that intelligently curate options rather than overwhelming users with exhaustive lists. This aligns with the principle of information overload, where users are overwhelmed by the volume and complexity of data presented to them.

But what is the process of making a choice? Schwartz's work highlights the dimensions of decision-making, illustrating the effort required to evaluate, compare, and ultimately finalize a choice. By automating certain aspects of this process, AI can alleviate the burden of decision-making and enable users to make informed choices more efficiently.

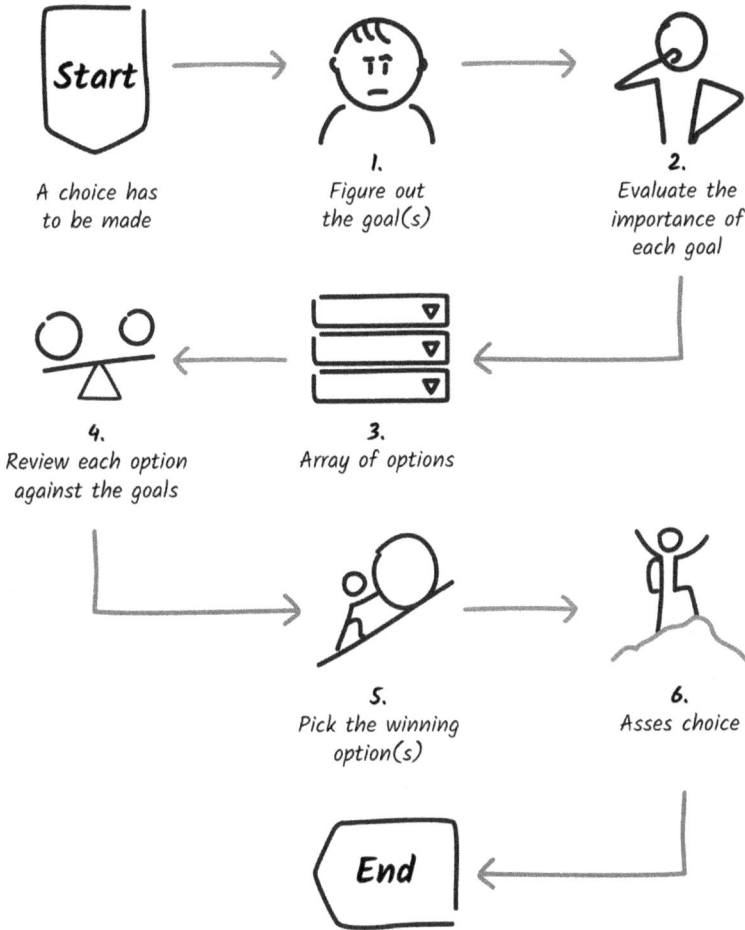

FIGURE 5.4 Steps required to make a decision and the corresponding cognitive effort, according to Schwartz.

5.2.1 Decision-Making Styles—*Maximizers* Versus *Satisficers*

Schwartz highlights two distinct decision-making styles that influence how people navigate choice: **maximizers**, who seek the absolute best option through exhaustive comparison, and **satisficers**, who aim for a solution that is "good enough" to meet their needs [19]. Satisficers' behavior is a practical approach, especially when exhaustive comparisons are costly or information is incomplete.

FIGURE 5.5   Visual illustration demonstrating framing bias in action by Scribbr [29].

Whether you are a maximizer or a satisficer, you will experience decision fatigue when confronted with many options. The key difference lies in your response to the process: satisficers are generally more content once a choice meets their standards, whereas maximizers often continue to evaluate alternatives, seeking an unattainable "perfect" option.

When discussing AI solutions, this presents an opportunity: by reducing the number of presented options, systems can satisfy both satisficers and maximizers, thereby preventing decision paralysis and improving overall satisfaction. For example, AI-driven solutions not only have the opportunity to simplify choices but also address cognitive biases that influence decision-making. For instance, **framing bias**—the way information is presented impacts decision-making. A study on packaging language demonstrates how subtle changes in phrasing influence user perceptions and choices [28]. AI systems can make a partial decision because these framing bias techniques do not affect them.

Another bias is the **contrast effect**. Comparing options relative to one another can distort perceptions, leading to suboptimal decisions. AI systems can help mitigate these biases by presenting information in a way that promotes rational, user-centric decision-making. By reducing the cognitive effort required to evaluate options, these systems can help users make more informed choices, minimize the psychological burden of regret and second-guessing, and increase overall satisfaction.

In conclusion, the psychology of choice reveals the profound impact that excessive options, cognitive biases, and decision fatigue have on user experience. AI technology can address these challenges by reducing the cognitive load associated with decision-making, aligning system outputs with user preferences and motivations, and mitigating biases through intelligent curation and presentation of information. By accommodating both maximizers and satisficers, AI services can simplify choices.

TL;DR

We live in a world of constant digital noise—notifications, choices, and decisions from the moment we wake up. Instead of simplifying life, technology often overwhelms us, leaving us mentally drained. This isn't a user problem; it's a **design problem**.

Drawing on the *Paradox of Choice by Schwartz*, this chapter examines how **excessive options**, particularly in digital interfaces such as Netflix or Amazon, contribute to **decision fatigue**, regret, and lower satisfaction. The more choices we have, the harder it becomes to decide—even on simple things like what to watch or buy.

I also explore **cognitive biases**, such as **framing effects** and the **contrast effect**, that further complicate our online decision-making process. People typically fall into two categories: **maximizers**, who seek the optimal solution, and **satisficers**, who are content with good-enough solutions. Both types suffer under the weight of excessive options.

This chapter introduces **anticipatory systems** as a way forward. Instead of forcing users to make constant micro-decisions, AI-powered systems can

- **Reduce cognitive overload** by surfacing only relevant options.
- **Automate low-stakes decisions** to make daily interactions smoother.
- **Help users make better choices** by presenting options in a clear and intuitive manner.

Ultimately, well-designed anticipatory systems don't just guess what users want—they relieve mental strain, respect autonomy, and support human well-being in an increasingly demanding digital world.

Origins of Anticipatory Design

6

What if design could help us think less, not more? In an age where every ping, swipe, and scroll demands our attention, the promise of anticipatory design isn't just convenience—it's cognitive relief. This chapter examines the origins of anticipation as a concept that predates digital technology, drawing on insights from biology, psychology, philosophy, and economics to demonstrate how anticipation has long served as a vital survival mechanism. It also reflects how the idea has evolved within digital services, laying the groundwork for the anticipatory systems we design today.

6.1 DEFINITION OF ANTICIPATION

Let's take a step back and consider a big question: what does it truly mean to anticipate something?

The word itself originates from the Latin *anticipalis*, meaning "to get ahead, foresee, or occupy beforehand." In many ways, that's precisely what good design should do—it should meet us one step ahead, offering assistance before we even ask. At its core, anticipation is about getting ahead—imagining what's coming and preparing for it. Whether it's your phone predicting where you're headed next or a health app nudging you to walk a little more today, these systems are built around the concept of using present-day data to forecast what might come next.

Anticipation isn't merely a new idea or a concept linked to technology; it's fundamentally connected to how humans (and even animals) survive. The ability to anticipate has been inherent in human thinking for centuries. In fields such as science, philosophy, biology, psychology, and economics, anticipation plays a central role in planning, acting, and adapting. It shapes behavior, influencing everything from how organisms respond to natural threats to how individuals make decisions in uncertain environments.

What happens when we incorporate this concept into design and technology? This book treats anticipation as more than a design tool; it's a way of thinking about intelligence, systems, and human experience. Anticipatory systems—whether in AI, user interfaces, or services—are built to do just that: act now based on what's likely to happen later. When done well, this kind of design doesn't just predict what users might want—it reduces mental strain, builds trust, and quietly supports better decisions.

But where do these capabilities come from? In anticipatory design, we use data and algorithms to predict user needs before they arise. However, we're not simply guessing—we must apply methods from fields like behavioral science, future studies, and human-centered AI to do it responsibly. This involves understanding how people think and make decisions, as well as what it takes to earn their trust. It also entails not overloading them with choices or removing too much control.

By the end of this chapter, you will understand not only the science behind anticipation but also the design mindset required to create the kinds of experiences that people genuinely desire from AI. First, however, let's take a step back and examine how various fields approach anticipation.

6.2 HOW FIELDS VIEW THE FUTURE DIFFERENTLY?

Each field offers its own perspective—some based on science, others on human behavior or philosophy—but collectively, they create the intellectual foundation of anticipatory design.

Biology
It views anticipation as crucial for survival. The biologist Robert Rosen famously argued that living organisms are inherently anticipatory—they continuously adjust their behavior in the present based on potential future events. This is not a luxury; it's a vital life skill.

Psychology
Reinterprets anticipation as prospection—the capacity to imagine and plan for future events. Research shows that we are not merely reactive to present stimuli; we actively pursue goals that lie ahead. The futures we envision shape the decisions we make today.

Sociology and Anthropology
They shift the focus from individuals to communities. Societies also create futures through rituals, narratives, aspirations, and political decisions. Anthropologist Arjun Appadurai reminds us that the future isn't merely an abstract idea; it's a cultural fact, something that communities build together.

Economics
Consider anticipation through the lens of expectations and incentives. Individuals make decisions today based on anticipated future benefits, be it investing in stocks or selecting

a career path. Economists analyze how these future-focused calculations influence current actions.

Design
Synthesizes all the information mentioned earlier into a tangible experience. In design, anticipation involves creating systems that respond before users take action, like suggesting a playlist before hitting play or highlighting a calendar conflict before scheduling a meeting.

6.3 THE SCIENCE BEHIND ANTICIPATION

If we want to build intelligent, forward-thinking systems, we need a clear understanding of what anticipation truly involves. Roberto Poli provides a valuable starting point [30]. He defines anticipation as "future-based information acting in the present." In other words, it's not merely about guessing what's next—it's about using future possibilities to influence our actions *now*.

Poli offers two complementary perspectives that help anchor this concept:

- A concise view: "Future-based information acting in the present."
- A more detailed explanation: anticipation involves a model of a future state, along with the ability to utilize that model to inform present-day decision-making.

This dual perspective—poetic and practical—highlights the key concept of anticipatory systems: they do not merely respond to current situations; instead, they proactively prepare for potential futures. When applied to design, this entails creating systems that are not simply reactive or responsive but proactive, silently shaping user experiences to align with needs that have yet to be fully articulated.

6.3.1 Anticipation Across Disciplines

Anticipation means different things in different fields, but it always plays a role in how we prepare for what's ahead. Understanding these different perspectives can help you draw from more profound knowledge when creating future-focused solutions.

Physics
In the 1940s, Italian mathematician Luigi Fantappiè introduced the concept of *syntropic processes*—a kind of phenomenon where waves converge and evolve forward in time, not caused by the past but shaped by future outcomes. These processes defy the usual flow of time and are now considered early examples of anticipatory behavior in natural systems.

Biology

In biology, anticipation plays a central role in how life functions. Organisms constantly adjust their behavior based on what *might* happen. This ability to act now in preparation for potential future conditions is considered essential to survival. From animals storing food to immune systems preparing for viruses, biology treats anticipation as a defining feature of life itself.

Psychology

Psychology sees behavior as driven more by goals than by immediate reactions to *stimuli*. Anticipation is tied to *prospection*—our capacity to imagine, plan, and prepare for future scenarios. People act not because of what the future *is* but based on how they mentally simulate what the future *could be*.

Sociology

Sociologists examine how societies manage time, particularly how they perceive and respond to the future. In today's world of uncertainty, sociology highlights the difference between distant futures and the "future-in-the-present," as well as the expectations, risks, and strategies that shape our lives today. Understanding time itself becomes essential to grasping anticipation at a social level.

Anthropology

While traditionally focused on the past, anthropology has increasingly studied how cultures imagine and shape their futures. It treats the future as a *cultural fact* constructed through imagination, anticipation, and aspiration. These elements define how people plan, hope, and act in the long term.

Economics

Economists often look to the future to explain present-day decisions. Unlike sociologists, who see the present as shaped by the past, economists assume people act today based on expected future rewards. Anticipation here is modeled through forecasts, probabilities, and decision theories that help predict behavior in uncertain markets.

Philosophy

Philosophy offers a more reflective perspective on anticipation, particularly in the realms of ethics and time. It explores how the past and future are not entirely separate from the present but are embedded within it. Philosophers examine how we act *in the now* with both memory and foresight, which shape our choices, even when outcomes are uncertain.

6.3.1.1 Design

Design treats anticipation as a way to *simplify* experiences. Anticipatory systems aim to respond to user needs before those needs are expressed—by predicting behavior, automating repetitive tasks, or making smart suggestions. These systems bridge past behavior and future intent, aiming to reduce friction and cognitive load.

Across disciplines, anticipation emerges as a cognitive and systemic tool for navigating uncertainty. It serves a dual function: guiding present actions toward desired futures and minimizing potential adverse outcomes [31]. These insights form the conceptual foundation for anticipatory systems, where predictive and future-oriented strategies are operationalized through technology. Our ability to anticipate empowers us to:

- Formulate novel goals and plans based on future needs.
- Develop increasingly abstract expectations to guide future actions and decisions.
- Mentally alter images before or in place of taking action physically.
- Significantly modify and adapt the environment to align with our goals and representations.

Each discipline offers a unique lens on how we approach the future—how we envision it, plan for it, and act upon it. Understanding this wide range of fields can be incredibly powerful. Whether you're building apps, services, or systems, you're not just solving today's problems; you're shaping what comes next.

Businesses should be more cautious about the costs and benefits of choice. There is a need to put more effort into understanding the implications of balancing the freedom of choice with the psychological burden that choice imposes. Younger, tech-savvy demographics, such as Millennials and Gen Z, are particularly vulnerable to these challenges. These groups are among the most digitally connected and frequently use platforms such as Netflix, Amazon, and social media, where cognitive overload is prevalent. Some studies and books, such as *The Anxious Generation* by Jonathan Haidt, have shown that high engagement with digital platforms is linked with increased stress and burnout [24]. These users often expect seamless and intuitive experiences but instead encounter systems that amplify their decision-making burden.

This is where the fields of behavior change, design, and anticipatory design can truly shine. If combined, these approaches can significantly reduce cognitive strain by helping our brains reduce the burden of complex mental calculations and mitigating the impact of negative emotional states.

Behavior change design empowers users to form better habits and make healthier choices. Anticipatory design, on the other hand, aims to anticipate decisions entirely by predicting users' needs and acting on their behalf. Together, these approaches may revolutionize how users interact with AI technology, transforming it from a source of stress into a tool that acts like a helpful partner rather than a demanding child.

For designers, the challenge is clear: how can we design AI-driven systems that not only understand human behavior but actively reduce cognitive strain? How can we balance automation with the agency to relieve users of decision fatigue without stripping users of control? These questions are at the heart of the UX discipline today as we struggle with the psychological consequences of the digital age.

6.4 EVOLUTION OF ANTICIPATORY DESIGN

As discussed in the previous chapter, anticipatory design has roots in both philosophy and psychology. It emerged as a solution to the growing problem of information overload. In the nineteenth century, philosopher Søren Kierkegaard described *angst* as the existential anxiety that comes from an overwhelming number of choices. In 2004, Barry Schwartz echoed Kierkegaard's sentiment in his book, *The Paradox of Choice*, arguing that too many options diminish decision-making capability and satisfaction [19]. This insight laid the foundation for anticipatory design, which seeks to alleviate the burden of choice by predicting and simplifying decisions before users are even aware of their needs.

6.4.1 Buckminster Fuller's Vision

In 1927, Buckminster Fuller, a visionary designer and architect, was the first to articulate the concept of anticipatory design. Fuller is often regarded as a man ahead of his time due to the inventions and research he conducted throughout his life. He consistently sought to anticipate the problems humanity would face and find technological solutions. One of the main objectives of his work, which he termed Comprehensive Anticipatory Design Science, is to achieve a better quality of life for everyone while using fewer resources.

This visionary framework encouraged designers to adopt a holistic approach, addressing complete systems rather than just individual components, and anticipating future challenges and trends while envisioning ideal outcomes [32].

In 1950, Fuller took a significant step toward realizing this vision by outlining the course of *Comprehensive Anticipatory Design Science*. Taught at MIT in 1956 as part of the Creative Engineering Laboratory, it stood out as one of its more unconventional offerings [32]. However, Fuller's aspirations for anticipatory design as a distinct field did not fully materialize [33]. Society may not have been ready for his visionary ideas at the time. Now, with the advent of the digital age—especially the rise of AI and intelligent systems—there's a renewed opportunity to put into practice the anticipatory design principles that Buckminster Fuller envisioned over 70 years ago. His ideas weren't just about solving isolated design problems; they were about reimagining how we shape the future through systems thinking, creativity, and foresight.

To grasp the depth of his vision, it helps unpack what he meant by "Comprehensive Anticipatory Design Science." Each word in the phrase carries a deliberate and layered meaning:

- **Comprehensive**: Encourages designers to frame the entire problem and work toward integrated solutions, promoting whole-system thinking rather than isolated fixes.
- **Anticipatory**: Involves identifying trends, understanding emerging challenges, and envisioning optimal outcomes before problems fully materialize.
- **Design**: Refers to the imaginative and practical process of shaping systems, tools, and experiences that respond to human and environmental needs.
- **Science**: Emphasizes the rigorous pursuit of understanding, using knowledge to address complexity with clarity, experimentation, and responsibility.

Building on this foundation, Aaron Shapiro introduced anticipatory design into the sphere of the digital era. In his 2015 article "The Next Big Thing in Design? Less Choice," [27] he argued that the future of user experience lies not in offering more options but in removing friction through intelligent simplification. Shapiro defined anticipatory design as "responding to needs one step ahead," positioning it as a way to reduce cognitive overload by automating low-stakes decisions. Aligned with the growing capabilities of AI, this approach aims to streamline digital experiences, making them not only more efficient but also more humane by anticipating user intent, eliminating unnecessary steps, and quietly assisting users in achieving their goals with less mental effort.

While Shapiro gave the concept a name and positioned it within the context of digital design, the seeds of anticipatory thinking had already been planted in earlier decades. Long before AI became mainstream, several companies were quietly laying the groundwork for what we now consider anticipatory experiences. These early systems may not have been familiar with the concepts of "cognitive load" or "intent modeling." Still, they shared a common goal: to simplify user decisions by making more intelligent, data-driven guesses about what might come next. Looking back at these milestones helps us understand how far the field has come—and what we've learned about designing systems that act with foresight rather than just reaction.

6.5 THE EVOLUTION OF ANTICIPATION IN DIGITAL PRODUCTS

The shift from personalization to anticipation didn't happen overnight. It evolved over decades through pioneering efforts by companies such as Amazon in the late 1990s, TiVo in the early 2000s, and Netflix shortly thereafter. Each played a key role in shaping what we now recognize as anticipatory experience design.

6.5.1 Late 1990s: Amazon's Recommendation Engine

Amazon revolutionized e-commerce by using purchase history to suggest products. Its "customers who bought this also bought" feature sets a benchmark for personalized recommendations, streamlining the shopping experience, and improving user satisfaction.

6.5.2 1999: TiVo's Predictive Recording

TiVo introduced the ability to learn users' viewing habits and automatically record shows of interest, marking one of the earliest applications of predictive technology in home entertainment.

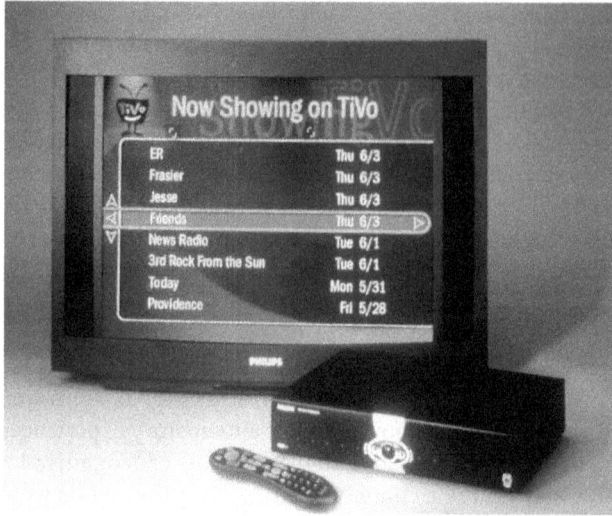

FIGURE 6.1   1999's TiVo.

FIGURE 6.2   Netflix in 2006.

Source: Web Design Museum [34]

6.5.3 2006: Netflix's Recommendation System

Netflix expanded on these ideas by introducing a recommendation engine that analyzed user ratings and viewing history to offer personalized movie suggestions. This system reduced the effort of searching through content, pioneering anticipatory design in the entertainment industry.

These early systems laid the foundation for what anticipatory systems have become today. Where personalization optimizes what is shown, anticipation predicts what should happen next, often automating part of the process. This shift enables technology to move from simply responding to user input to preemptively solving problems, reducing friction, and fostering a sense of effortless interaction.

The evolution from personalization to anticipation reflects a natural progression in digital design, where data-driven insights are increasingly used to predict and fulfill user needs. Early systems, such as Amazon's recommendations and TiVo's predictive recordings, demonstrated the value of personalization, while Netflix's recommendation engine showcased the potential for anticipatory systems to streamline user experiences. Today, as AI gains greater predictive power, anticipatory systems have the opportunity to move beyond personalization, offering proactive, user-centered solutions that simplify decision-making and improve satisfaction.

TL;DR

What is anticipation?—Anticipation is the act of using imagined futures to inform present decisions. It's not about prediction for prediction's sake—it's about using future-facing logic to reduce complexity, support decision-making, and align technology with human behavior. In design, it's the difference between waiting for user input and proactively supporting user intent.

Anticipation across disciplines—Anticipation is a core feature of intelligent behavior across domains.

- **Biology** treats it as a survival mechanism.
- **Psychology** sees it as prospection. A mental time travel that guides action.
- **Sociology and anthropology** explore how communities build collective futures.
- **Economics** models it as expectation-driven behavior.
- **Philosophy** ties it to ethics, agency, and the experience of time.
- **Design** operationalizes all of this to reduce friction and offer meaningful support before users make a request.

The Rise of Anticipatory Design—Modern life is overloaded with content, options, and cognitive demands. From streaming platforms to digital assistants, users are increasingly fatigued by the sheer number of choices and complexity. Anticipatory design emerged to address this overload by minimizing unnecessary decisions and framing better ones. Inspired by thinkers such as Søren Kierkegaard and Barry Schwartz, and popularized by Aaron Shapiro, it offers a new model: technology that acts one step ahead to support, rather than control, our decisions.

Buckminster Fuller envisioned anticipatory design as a whole-systems approach long before the digital era—a vision that has regained urgency in the age of AI.

From Personalization to Anticipation—Early personalized systems, such as Amazon and Netflix, laid the groundwork by responding to past behavior. Anticipatory systems go further, combining historical data with real-time context to act proactively before users request it. However, with this power comes risk: poor implementation can feel invasive or reduce user autonomy. The challenge is to balance intelligent automation with human agency—designing systems that anticipate needs without compromising trust, transparency, or freedom of choice.

The Fragile Promise of Anticipatory Experiences

7

In the previous chapters, we traced the origins of anticipatory design—from its theoretical foundations to its early expressions in digital products. Now, let's shift focus to the present. While the vision of anticipatory design remains compelling, its real-world implementation reveals a more turbulent landscape. The gap between ambition and execution has become increasingly evident, raising essential questions about why so many services fail to deliver on the promise of acting one step ahead and helping alleviate users' cognitive overload.

Based on my observations across various industries, those who focus on creating anticipatory experiences gain significant benefits. Services now personalize content, proactively guide users, and support decision-making in increasingly intelligent ways. The promise is clear: reduce cognitive overload, anticipate needs, and deliver more meaningful outcomes with less effort.

In finance, these systems promote habit-formation and long-term savings. In education, they tailor learning paths and reduce frustration. In transportation, they optimize routes and encourage safer travel. In healthcare, they foster preventive care and behavioral change. In retail, they enhance customer satisfaction while reducing costly returns. Despite these successes, one key insight stands out: personalization falls short without context and adaptability.

These experiences are most effective when they are dynamic, context-aware, and aligned with evolving user needs, rather than relying solely on historical data. While the benefits are compelling, the failures that follow show what happens when anticipatory design lacks nuance, transparency, or emotional intelligence.

DOI: 10.1201/9781003642800-9

7.1 THE PRESENT TENSE OF ANTICIPATORY DESIGN

Anticipatory design has emerged as a promising design paradigm in response to the growing cognitive strain of modern digital life. It not only holds great potential but also raises critical questions: how can anticipatory systems enhance efficiency and satisfaction while preserving user control and autonomy?

To understand how anticipatory design differs from personalization, consider this analogy: a waiter at a restaurant. Traditionally, the waiter waits for you to order. A personalized system might remember your usual drink. However, an anticipatory system goes one step further—it recognizes subtle cues, such as your consistent choice of sparkling water, and offers it before you even ask. Anticipation combines historical data with real-time context, facilitating informed decision-making and enhancing seamless experiences.

This shift—from personalization to proactive intelligence—is not just about convenience. It's about reducing friction, minimizing cognitive biases, and supporting better decision-making by aligning with what users genuinely need, often before they articulate it. Done well, anticipatory systems reduce information overload, mitigate decision fatigue, and adapt to individual styles—whether users are satisficers seeking quick solutions or maximizers who want to weigh every option.

As AI systems become increasingly sophisticated, user expectations have also evolved. People now expect digital experiences that go beyond personalization and move toward proactive, context-aware assistance. This shift frames the current state of anticipatory design, setting the stage for both its most promising developments and its most pressing challenges.

Although Amazon, TiVo, and Netflix are among the earliest examples of anticipatory experiences, the **Nest smart thermostat** follows closely behind as one of the first connected devices to apply anticipatory principles in the home. Launched in 2011 by former Apple engineers Tony Fadell and Matt Rogers, Nest quickly gained attention for its ability to learn from historical user behavior, weather patterns, and occupancy trends, thereby adjusting heating and cooling proactively. It might warm your home on a cold morning before you even think to touch the thermostat—or dial back energy use when it senses no one is home. In 2014, Google acquired Nest for $3.2 billion, signaling a significant investment in the future of intelligent, proactive home systems. Today, Nest remains a market leader in smart home technology, demonstrating how anticipatory design can transition from novelty to mainstream success. It turned what was once a purely reactive object into a proactive, personalized experience—subtly aligning comfort with context.

But even systems as sophisticated as Nest have limits. Imagine a household with a newborn baby. Comfort and safety now outweigh energy savings; yet, the thermostat, unaware of this context, might continue to prioritize efficiency. This is where **automated systems can fall short**: when the prediction makes sense *statistically* but fails to account for human nuance. It's a reminder that anticipatory systems need more than

FIGURE 7.1   Nest thermostats by Google.

good data—they need a design that understands people. To ensure that these systems empower rather than frustrate, designers must focus on three key principles:

- **Transparency and trust**: Users should understand how predictions are made and feel they can override or customize them.
- **Adaptability**: Systems must evolve with users, learning from new patterns and refining predictions over time.
- **User-centric outcomes**: Predictions should focus on user goals rather than solely on business metrics to prevent the automation of irrelevant or poorly timed actions.

When these principles are established, anticipatory systems can feel intuitive, empowering, and even delightful. Conversely, when they are not implemented, there is a risk of eroding trust or creating experiences that feel uncanny, opaque, or misaligned with users' actual needs. If not, poorly executed anticipation can alienate users instead of empowering them. Inaccurate predictions feel intrusive and over-automation risks stripping users of control. Overly aggressive filtering can confine users within "invisible filter bubbles," limiting their discovery, agency, and diversity of experience. That's why anticipatory systems must be more than smart—they must be transparent, adaptive, and aligned with users' evolving goals and values. Only then can they fulfill their promise of not just anticipating needs but truly supporting them.

7.1.1 Case Study: IBM's Watson for Oncology

The IBM Watson for Oncology project highlights the potential pitfalls of AI. The project was canceled after $62 million in investments due to unsafe treatment recommendations.

Initiated in 2013 in collaboration with the University of Texas MD Anderson Cancer Center, the project aimed to develop an AI-driven oncological advisor to recommend personalized treatments based on clinical guidelines. However, in July 2018, StartNews [35] reviewed internal IBM documents and revealed that Watson made erroneous and dangerous recommendations due to reliance on a small set of hypothetical patient cases during training. These issues were compounded by a lack of scalability, transparency, and user trust in its decision-making process. Although functional, the user interface did not address transparency in recommendations, leaving health professionals unable to validate or trust the system's outputs. The interface failed to provide the necessary explainability to instill user confidence.

The Watson for Oncology case underscores the importance of transparency and explainability in any AI-driven system. To foster trust, systems must explain their predictions, recommendations, and anticipations in ways that users can understand and evaluate. Simply achieving usability is insufficient; trust depends on users perceiving the system as transparent and aligned with their goals. Building trust also involves managing user expectations around the capabilities and limitations of anticipatory systems. Since AI operates based on probabilistic models, users often overestimate its capabilities. Clear explanations can help users understand when to trust the system and when to rely on their own judgment. This balance is critical for minimizing user frustration and fostering long-term trust as both AI technologies and user relationships evolve.

7.2 OVERPROMISING AND UNDER-DELIVERING: PITFALLS OF ANTICIPATORY DESIGN

As businesses increasingly embrace anticipatory design, the ability to proactively address user needs while aligning with long-term objectives becomes critical. This dual focus ensures that immediate actions cater to user expectations while supporting broader strategic goals. Despite its promise, anticipatory design often fails in practice, with few notable successes. The following cases—**Digit**, LifeBEAM **Vi Sense** Headphones, and **Mint**—underscore significant shortcomings in execution, offering valuable insights for future systems. Each example highlights how a lack of user behavior comprehension and poorly implemented predictive capabilities resulted in a disconnect between user expectations and service delivery.

7.2.1 Digit: Struggling With Contextual Understanding

Digit was designed to anticipate and automate personal savings by analyzing user spending habits and transferring small amounts into savings accounts. While the concept was

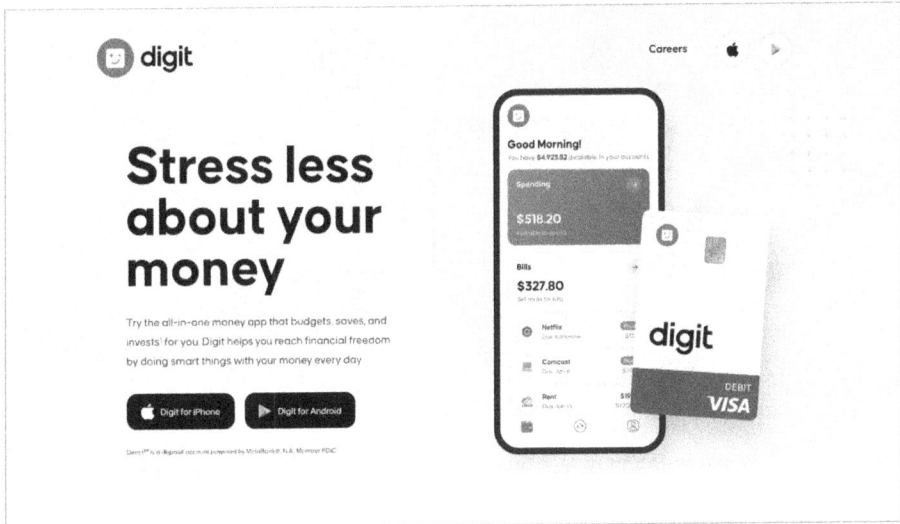

FIGURE 7.2 Screenshot of Digit app website as it appeared in 2020.

innovative, the system's failure to incorporate real-time contextual understanding led to a poor user experience. Unexpected withdrawals created financial stress for users living paycheck to paycheck, ultimately breaking trust in the system.

For Digit, a broader understanding of user behavior and financial context could have been integrated through foresight techniques like *scenario planning* and *horizon scanning*. These methods would have anticipated edge cases, such as users with inconsistent cash flow or emergencies, ensuring the service adapts to diverse user financial conditions. Digit relied heavily on historical spending patterns without dynamically adjusting to unforeseen financial events or changing user needs. Incorporating real-time analytics with predictive algorithms could have mitigated these issues by refining forecasts based on evolving user data.

Digit's lesson learned is that anticipatory systems must integrate contextual awareness to adapt to user behaviors and unforeseen circumstances dynamically. Transparency in decision-making processes is crucial for building and sustaining user trust.

7.2.2 LifeBEAM Vi Sense Headphones: Complexity and User Experience Challenges

Launched in 2016, LifeBEAM's Vi Sense headphones were marketed as the first AI-powered fitness coach, combining voice-based interaction with real-time biometric feedback. The product promised more than just fitness tracking. It envisioned a personal trainer in your ears, capable of offering dynamic, personalized coaching through conversational AI. But in practice, Vi struggled to meet these expectations. The system only supported outdoor running, excluded other fitness modes such as cycling or

FIGURE 7.3   Representation of Vi's headphones and application as it appeared in 2018.

indoor training, and offered just six predefined goals, which were insufficient for more experienced users. Rather than adapting to user progress or varying contexts, the AI delivered static and unresponsive coaching. This disconnect between the vision of truly adaptive AI and the product's rigid functionality ultimately left many users frustrated and disengaged.

Strategic foresight methodologies, such as *trend analysis* and the *Delphi method*, could have helped the design team anticipate potential user pain points, including the need for adaptive fitness coaching. By envisioning diverse user scenarios—ranging from novice users to experienced athletes—the team could have ensured that the product met varying expectations.

Predictive modeling based on fitness data could have enabled a more nuanced understanding of individual user behaviors, aligning workout recommendations with personal goals and progress. Regular feedback loops between the user and system would have refined the predictive algorithms over time. LifeBEAM Vi Sense's lesson learned is that personalization must be a cornerstone of anticipatory systems, ensuring that the experience evolves in response to individual needs and motivations. Additionally, transparency and clarity in system interactions foster user engagement and adoption.

7.2.3 Mint: Misalignment With User Goals

Mint aimed to empower users to manage their finances by providing automated budgeting tools and financial advice. While the service had the potential to anticipate user needs, users often found that the suggestions were not tailored to their unique financial situations, resulting in generic advice that did not align with their personal goals. The lack of personalized, actionable steps led to a mismatch between user expectations and service delivery. This misalignment caused some users to disengage, as they felt that Mint was not fully attuned to their unique financial journeys.

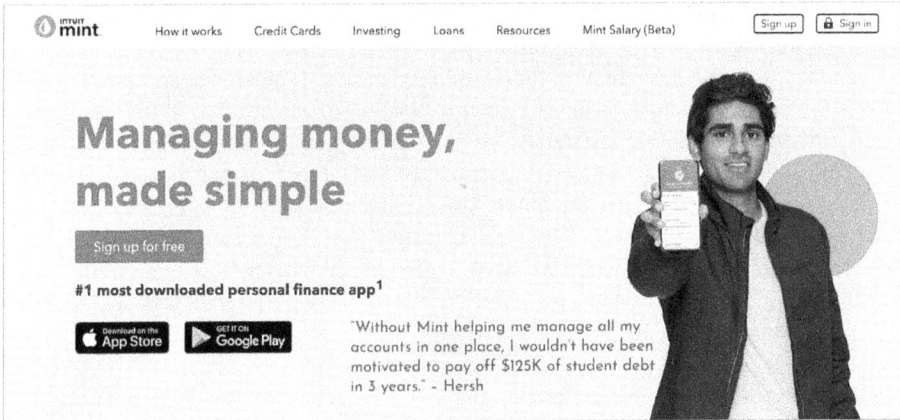

FIGURE 7.4   Screenshot of the Mint app website as it appeared in 2021.

By employing *visioning planning,* and *scenario development,* Mint could have identified diverse user financial archetypes and tailored its advice to different needs. Understanding user aspirations and concerns would have allowed Mint to create more meaningful, goal-oriented recommendations.

Mint's reliance on rigid financial models without adapting to dynamic user inputs undermined its potential. By integrating real-time financial data and leveraging ML for predictive insights, the system could have aligned its recommendations more closely with individual goals and motivations. Mint's lesson learned is that to ensure relevance, anticipatory systems must align forecasts with user aspirations and offer actionable, personalized insights. Systems that fail to adapt risk alienating their users and losing engagement.

7.2.4 Additional Honorable Mentions

While Digit, Mint, and Vi Sense stand out as clear cautionary tales, several other anticipatory services offer additional lessons worth highlighting:

Google Now was a bold attempt to deliver proactive assistance via context-aware notifications. While groundbreaking in its ambition, users struggled with its lack of transparency. The app surfaced information without clear logic or user control, leading many to perceive it as invasive rather than helpful. Eventually, Google folded it into the more flexible Google Assistant.

Clippy, Microsoft's infamous assistant, was an early—and often mocked—experiment in anticipatory user interface. While it aimed to help users complete tasks proactively, its interruptions were frequently mistimed and unhelpful. It assumed too much based on too little context, becoming a symbol of intrusive design.

Amazon's dash buttons introduced a new form of anticipatory commerce by allowing users to reorder household goods with a single press. While innovative, they underestimated the variability in user preferences and price sensitivity, ultimately failing to justify their existence in a world increasingly driven by dynamic interfaces and smarter shopping tools.

Hello sense was a sleek sleep tracker that promised to personalize users' rest by tracking sleep cycles and environmental factors. Despite its elegant design, it failed to deliver clear, actionable insights that users could trust or apply. Launched in 2014, Hello Sense struggled to retain users due to its failure to provide clear, actionable insights. Despite its elegant design, it was eventually discontinued in 2017.

These examples reinforce a recurring pattern: systems that assume too much, communicate too little, or fail to adapt to individual needs often fall short of delivering meaningful value. Even the best algorithms can fail when they ignore human nuance, trust, and transparency. These "honorary mentions" remind us that the road to anticipatory design is paved with good intentions but only thoughtful execution leads to sustained success.

7.3 LESSONS LEARNED

Digit, LifeBEAM Vi Sense, and Mint's experiences reveal a common pitfall in anticipatory design: overpromising and under-delivering. While aiming to simplify our lives, these services often fail to account for the complexity of human behavior.

For instance, Digit, with its automated savings, frustrated users with opaque and unpredictable decisions. LifeBEAM Vi Sense, despite its ambitious goals, fell short of expectations by failing to provide sufficient customization options for both novice and experienced users, leading to frustration and disengagement. And Mint, despite its ambition, offered rigid budgeting tools that fell short of providing truly personalized financial insights. These examples underscore the critical need to strike a balance between proactive assistance and user control.

These failed services reveal a deeper tension in anticipatory design: the gap between what systems predict and what people actually need. Much of this gap stems from an overreliance on **forecasting**—data-driven models designed to extrapolate from the past—without the balancing perspective of **foresight**, which accounts for uncertainty, context, and emerging needs. In each of the cases examined, this imbalance led to missed signals, fragile assumptions, and experiences that failed to resonate in real life.

Although I will examine these two methods more thoroughly later in the book, it's crucial to mention them here as essential tools for creating anticipatory experiences that align with genuine human goals, rather than missing the mark.

Therefore, to successfully tackle these issues, effective anticipatory systems must incorporate foresight and forecasting principles to:

- **Adapt to evolving needs**: Mint and Digit highlight the risks of static systems that fail to adapt to users' changing contexts. A foresight-driven approach ensures that systems remain flexible and responsive to emerging behavioral patterns, not just past ones.
- **Prioritize user-friendliness**: LifeBEAM Vi Sense illustrates how overly complex systems can alienate users. Forecast-driven simplicity and a foresight-informed understanding of diverse user scenarios can strike a balance between functionality and accessibility.
- **Build trust**: A lack of transparency in decision-making processes eroded trust across all three cases. Anticipatory systems must not only forecast outcomes but also clearly frame their reasoning, supporting autonomy and explainability.
- **Anticipate the unexpected**: As Digit demonstrated, forecasting alone often struggles to account for outlier events. Foresight methodologies like scenario planning and environmental scanning help surface edge cases and prepare systems for irregular human contexts.

Instead of simply reacting to data, these systems must proactively consider diverse user scenarios and adapt accordingly. This requires a blend of data-driven forecasting and a broader, more strategic approach considering human behavior and values.

Anticipatory systems have demonstrated immense potential when implemented with nuance across various industries, including finance, education, healthcare, and others. They drive habit formation, personalize pathways, and support proactive decision-making. But these benefits can only be sustained if systems are designed to evolve alongside users, not just predict them. Without frameworks that balance probabilistic forecasting with strategic foresight, anticipatory experiences will continue to promise more than they can deliver.

By embracing this holistic perspective, we can develop anticipatory systems that genuinely enhance our lives, rather than frustrating or overwhelming us. This involves:

- **Dynamic contextual awareness**: Ensuring systems adapt in real time to changing user conditions.
- **Seamless personalization**: Moving beyond generic responses to deliver meaningful, tailored interactions.
- **Transparent decision-making**: Communicating system logic to foster trust and user confidence.

In conclusion, the failures of these services underscore the need for anticipatory systems to combine the data-driven precision of forecasting with the strategic breadth of foresight. By leveraging foresight to envision diverse user scenarios and employing forecasting to predict immediate needs, anticipatory systems can deliver meaningful, adaptive, and user-centered experiences. This alignment of foresight and forecast is not just a best practice—it is necessary for businesses seeking to build trust, foster engagement, and navigate the complexities of user expectations in a rapidly evolving landscape.

Ultimately, anticipatory design's true promise isn't merely prediction—it's alignment. It's about creating systems that act not only for people but also with them, constructing experiences that feel intelligent because they are based on empathy, context, and intent.

7.4 WHERE DESIGN MAKES A DIFFERENCE?

The challenges faced by anticipatory systems aren't just technical—they're profoundly human. At their core, successful anticipatory experiences are fundamentally about behavior. Anticipation involves more than data-driven predictions; it requires understanding how people think, decide, and interact with the world around them. That means moving beyond the illusion of seamless automation and acknowledging the importance of user **agency, trust, and control** within intelligent systems.

Trust is the foundation of any anticipatory system. Users need clarity about how these systems function, reassurance that their autonomy is respected, and confidence that the system genuinely aligns with their interests. Without that, even the most sophisticated predictions risk falling flat. To build trust, designers must understand how people perceive control, how they evaluate transparency, and how they negotiate shared agency with machines.

Anticipatory systems must thus integrate human-centered principles to fulfill their promise effectively. These principles include the following:

- **Respecting autonomy**: Recognizing that users value personal control and the freedom to intervene, modify, or override automated behaviors when needed.
- **Enhancing transparency**: Clearly communicating how anticipations and predictions are generated, aligning explanations with human expectations and cognitive patterns.
- **Empowering users**: Balancing proactive assistance with meaningful options for independent decision-making.

When these behavioral principles are thoughtfully applied, anticipatory systems can deliver genuinely transformative outcomes:

- **Reducing cognitive load** by surfacing only what matters most.
- **Alleviating decision fatigue** by automating the right choices at the right times.

- **Facilitating better decisions** by making information more usable and accessible.
- **Increasing satisfaction and efficiency** by streamlining interactions in ways that align with users' goals and mental models.

Yet as we've seen, human behavior is not easily predicted. It's fluid, contextual, and influenced by far more than data can capture. Anticipatory design must reflect this complexity, rather than flattening it. Systems that reduce users to past behavior patterns risk undermining trust and usability. The real strength of anticipatory design lies in honoring variability and designing for agency, not just efficiency.

That's why, before diving into the science of behavior change, we must first explore a more foundational tension: how to balance autonomy and automation in systems that act on behalf of users? The next chapter tackles this head-on, offering a framework for navigating shared control, designing with transparency, and creating intelligent systems that act not only *for* people but also *with* them.

TL;DR

Anticipatory design holds great potential—but its real-world implementation is riddled with challenges. This chapter examines the current landscape of anticipatory experiences, highlighting both their promise and their pitfalls. While early innovations, such as Nest, demonstrate how AI-powered systems can personalize and proactively support users, many modern services still fall short. Case studies such as IBM Watson for Oncology, Digit, Vi Sense, and Mint reveal a common thread: a lack of transparency, contextual sensitivity, and adaptability leads to broken trust and disengagement.

To succeed, anticipatory systems must go beyond historical data. They need to be:

- **Transparent**: Clearly communicate how decisions are made.
- **Adaptable**: Evolve with users' changing goals and contexts.
- **User-centered**: Align predictions with fundamental human needs, not just business metrics.

By combining **forecasting (data-driven predictions)** with **foresight (strategic imagination)**, designers can build experiences that are both intelligent and humane. The future of anticipatory design lies not in more automation but in more empathy, trust, and alignment with what truly matters to people.

Trust Calibration
Balancing Autonomy and Automation

8

The previous chapter helped us understand that as anticipatory systems grow in autonomy, so do the risks of over-automation, reduced user agency, and mismatched expectations. This chapter examines the design of AI systems that strike a balance between automation and human control, emphasizing the importance of explainability, transparency, and shared responsibility. It introduces models for aligning automation levels with appropriate explanation depth and maps the evolving dynamics of human–AI collaboration.

Anticipatory design promises seamless, proactive experiences. But in practice, this vision can backfire. When systems act *too much* on a user's behalf, they risk automating away the very things people care about: autonomy, agency, and meaningful choice. As we've seen in the previous chapter, poorly calibrated automation often leads to mismatched expectations, user frustration, and a slow erosion of trust.

A core challenge lies in the disconnect between how systems behave and how users *expect* them to act. Mental models—how users internally understand and predict system behavior—play a critical role in shaping trust. (I'll explore mental models more deeply in an upcoming chapter.) For now, it's essential to recognize that explainability, feedback, and appropriate levels of user control are key to aligning these models with system functionality.

Designing trustworthy AI is not just a matter of ethics or transparency—it's about experience. When people understand how a system works and feel they can influence it, they're more likely to use it confidently. But when systems are opaque or overly prescriptive, users feel sidelined, leading to disengagement or outright rejection.

One of the most concerning manifestations of this is the **"out-of-the-loop" effect** [36]. This effect occurs when users become so detached from the decision-making process that they lose situational awareness and the ability to step in or intervene when needed. Initially identified in high-stakes fields such as aviation and military systems, the out-of-the-loop effect is becoming increasingly relevant to anticipatory systems that automate everyday tasks. When users are not adequately informed or involved, they may either blindly trust or mistrust the system entirely, without fully understanding its logic or limitations.

To mitigate this, designers must address two interdependent aspects: **explainability** and **user control** [37]. Clear, context-sensitive explanations help users form accurate expectations, while well-designed controls allow users to override, adjust, or even question automated decisions. Together, these elements foster a sense of shared agency, positioning the AI not as a black box or an authoritarian force but as a cooperative partner.

Ultimately, trust in anticipatory systems isn't granted—it's calibrated. That calibration depends on how well we balance automation with human autonomy, clarity with control, and prediction with participation.

8.1 AUGMENTATION VERSUS AUTOMATION

In today's AI-driven experience economy, one of the first design questions is whether the system should emphasize augmentation or automation. This distinction is foundational to anticipatory design.

Automation refers to machines taking over a task, while **augmentation** means AI supports human performance without removing the human role. These two models often coexist and reinforce one another, but they must be carefully balanced. For example, in customer service, an automated chatbot can handle frequently asked questions (automation), freeing up human agents to focus on more complex issues where their empathy and problem-solving skills are crucial (augmentation). In practice, augmentation and automation are often intertwined. These dual AI applications coexist across time and context, creating a paradoxical tension [38]. When designed thoughtfully, they can complement each other by simplifying and enhancing the outcomes of complex processes.

A core principle of anticipatory design is reducing decision fatigue. While fewer choices may relieve mental strain, users often still want the **freedom to explore** and make informed decisions. But removing too many options can feel disempowering. Simplifying too much can create a sense of manipulation, eroding trust and agency.

Consider the example of a fitness app that recommends workouts based on an individual's past behavior. While helpful for some, others might want to choose routines based on their current mood or health condition. Systems that assume too much risk alienate users. To support both autonomy and automation, we must offer explainability and flexibility by **striking a balance between automation and user agency**.

Examining frameworks like the *Ten Levels of Automation* is a helpful way to understand how anticipatory systems strike a balance between automation and user control [39]. These models outline the spectrum of system autonomy, from complete human control to full machine autonomy. However, while authors like Kaushik et al. mapped these automation levels in technical terms, their framework does not deeply consider the evolving dynamics of **human–AI collaboration** or the varying **trust thresholds** users experience at each stage. For instance, their model focuses mainly on delegation of control and response time, without sufficiently addressing how user roles change or how trust must be actively designed across different levels of autonomy. Another author, Kore, builds on this by introducing trust into the equation in his book *Designing Human-Centric AI Experiences*. Yet his work also fails to capture the nuances of shifting human roles and responsibilities in increasingly autonomous systems [37].

To address these gaps, this book expands on their work by mapping system autonomy alongside the evolving roles of humans and AI, how responsibilities are shared, and how trust must be recalibrated at each level. The following section presents this expanded model, providing a more comprehensive view of how anticipatory systems can facilitate dynamic and trustworthy collaboration.

8.1.1 Levels of Automation and Explanation

The degree of explanation required to build and sustain trust in AI systems is tightly connected to the system's level of automation [37]. As automation grows, the need for clear communication about how these systems work also increases. A general overview is usually sufficient for simpler systems, such as voice assistants. However, for complex systems, such as those involving medical aides or financial decisions, it's essential to provide detailed explanations about where the data originate, how decisions are made, their limitations, and potential risks. This clarity helps users understand the system's reliability and maintain control.

As automation advances, the need for transparency in design becomes more critical. For instance, a rudimentary recommendation engine suggesting products based on a simple purchase history might only require a concise explanation, such as "Items similar to your past orders." However, consider a fully automated loan application system used by a financial institution. Such a system, which makes critical decisions about individuals' financial futures, demands far greater transparency. Applicants denied a loan need a clear understanding of *why* their application was rejected. This necessitates detailed information regarding the specific data points considered (e.g., credit score, debt-to-income ratio), the algorithms used to assess risk, the measures implemented to mitigate potential biases in the evaluation process, and any known limitations of the system's predictive capabilities. This level of transparency is essential to ensure fairness, accuracy, and accountability in high-stakes automated decision-making. Therefore, Table 8.1 outlines the different levels of automation, along with the corresponding depth of explanation required for each.

Therefore, the level of automation in a system directly affects the kind of explanation needed to maintain user trust. As automation increases, the demand for detailed, transparent, and context-aware communication increases.

TABLE 8.1   Balancing Automation and Explanation to Build Trust in AI Systems.

LEVEL OF AUTOMATION	LEVEL OF EXPLANATION	DESCRIPTION	EXAMPLES
Low The AI performs simple, noncritical tasks that require frequent human intervention.	**Basic** General overview of AI factors.	Users receive a high-level overview of how the AI works without going into technical details.	Voice assistants, recommendation systems.
Moderate The AI handles more complex tasks with significant autonomy, albeit not completely. Users might interact frequently with the AI for guidance or assistance.	**Moderate** Clear explanations of how the system makes decisions, key features, and main limitations.	Users are provided with clear, understandable explanations of the AI's decision-making processes, including key features and limitations.	Automated customer service (bots), navigation tools.
High The AI supports critical decision-making processes, but final decisions often involve human judgment.	**Thorough** Detailed descriptions of decision-making processes, including model operation, data used, and limitations.	Detailed information about how the AI model functions, including its data sources, methodologies, and constraints. Users need to understand how the system arrives at its recommendations.	Medical decision tools, financial AI advisors.
Full AI operates autonomously with minimal to no human intervention required in real time.	**Comprehensive** Nearly fully descriptive operation, exact limitations, and robust transparency in critical and noncritical scenarios.	Comprehensive understanding of the AI's operation, covering nearly all aspects of its decision-making process, exact limitations, and potential failure modes. Essential for trust and safety.	Self-driving cars, autonomous drones, criminal justice AI.

As a designer, use this table during design sprints or product discovery sessions to ask:

- What level of automation are we targeting in this feature?
- Are we providing **enough explanation** for the user to trust this level of autonomy?
- Should we surface more transparency, or offer manual override controls?

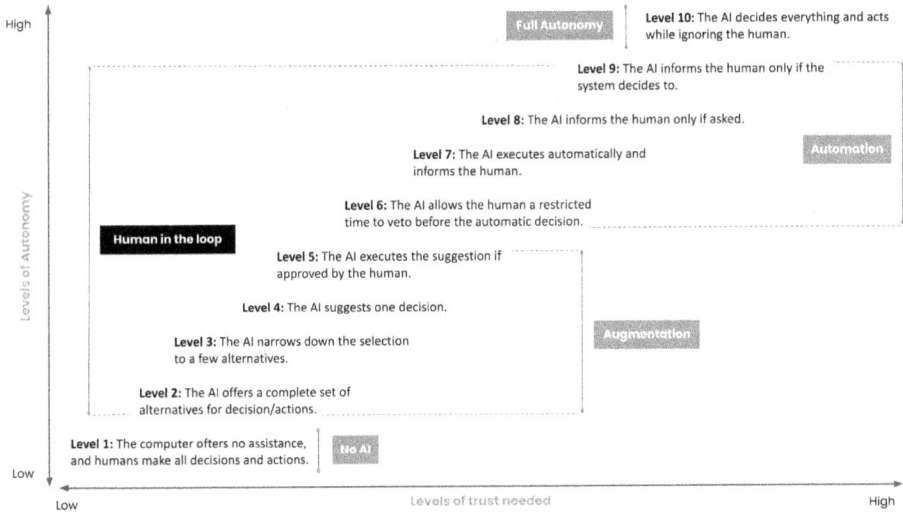

High

Level 10: The AI decides everything and acts while ignoring the human.

Full Autonomy

Level 9: The AI informs the human only if the system decides to.

Level 8: The AI informs the human only if asked.

Level 7: The AI executes automatically and informs the human.

Automation

Level 6: The AI allows the human a restricted time to veto before the automatic decision.

Human in the loop

Level 5: The AI executes the suggestion if approved by the human.

Level 4: The AI suggests one decision.

Level 3: The AI narrows down the selection to a few alternatives.

Augmentation

Level 2: The AI offers a complete set of alternatives for decision/actions.

Level 1: The computer offters no assistance, and humans make all decisions and actions.

No AI

Levels of Autonomy

Low

Low   Levels of trust needed   High

FIGURE 8.1   Levels of system autonomy according to Kore [37].

The following figure expands on this. It presents a spectrum of *autonomy levels* (on the vertical axis) against the *levels of trust required* (on the horizontal axis). It categorizes automation into stages ranging from low automation, where humans are entirely in control, to full automation, where the AI operates independently with minimal or no human intervention.

- Levels 1–2 represent systems without AI involvement, where humans make all decisions and take full responsibility. No trust in AI is required.
- Levels 3–6 are those systems beginning to assist through **augmentation**, suggesting or narrowing decisions while leaving final choices to the user. These *HITL* stages require moderate levels of explanation to foster understanding and trust.
- Levels 7–10 represent **full autonomy**, where AI acts independently and may inform users only selectively. These systems must offer **comprehensive, context-sensitive explanations** to maintain transparency and validate trust. Here, user control calibration becomes essential—not only for quality of experience but also for safety and accountability.

In summary, as systems evolve from decision-support tools (Levels 2–3) to full autonomy (Levels 7–10), the user's role transitions from an active decision-maker to an occasional intervener or monitor. Mid-range levels (3–6) emphasize HITL collaboration, requiring clear explanations and validation mechanisms. Higher levels, in contrast, must rely on strong transparency, adaptive interfaces, and clear fallback options to maintain trust in the system's independent decision-making.

8.2 RETHINKING HUMAN-AI COLLABORATION

Kore and Kaushik's models provide a solid foundation, yet they lack a clear connection of how shared responsibility changes with increasing autonomy. Thus, the following table assesses the need to include a dimension that outlines the shifting roles and responsibilities of humans and AI at various levels.

These stages make clear that **trust calibration** is not static—it evolves alongside the user's role, the AI's capabilities, and the interface that mediates between them. This

TABLE 8.2   System Autonomy Versus Human Input [40].

LVL	DESCRIPTION	HUMAN ROLE	AI ROLE	SHARED RESPONSIBILITY	LEVEL OF TRUST NEEDED
1	No AI: Humans make all decisions.	Full decision-making and execution.	None	None	None (No AI involvement).
2	AI offers a complete set of alternatives.	Evaluate all options and make the decision.	Generate and present all possible alternatives.	Information gathering and presentation.	Low: Human retains full control.
3	AI narrows down a few alternatives.	Select the best option from AI's suggestions.	Filter and refine the list of alternatives.	Decision refinement and risk evaluation.	Low-moderate: Trust in AI's filtering ability.
4	AI suggests one decision.	Evaluate the suggestion and decide to act.	Propose the most optimal single decision.	Validation of AI suggestions.	Moderate: Trust in AI's ability to identify the best option.
5	AI executes with human approval.	Approve or reject the AI's proposed action.	Execute the action once approval is given.	Agreement on execution criteria.	Moderate-high: Trust in AI's execution capabilities.
6	AI allows a veto before an automatic decision.	Monitor AI decisions and veto if necessary.	Execute decisions automatically after a delay.	Define veto criteria and monitoring feedback.	High: Trust in AI's ability to act without human interference.
7	AI executes automatically and informs the human.	Set parameters for AI decisions; monitor outcomes.	Execute actions autonomously and notify of results.	Ensure decision criteria are met.	High: Trust in both execution and reporting.

TABLE 8.2 continued

LVL	DESCRIPTION	HUMAN ROLE	AI ROLE	SHARED RESPONSIBILITY	LEVEL OF TRUST NEEDED
8	AI informs humans only if asked.	Rarely involved; intervene only on request.	Autonomously act and store justification data.	Define edge cases for human intervention.	Very high: Trust that AI can act without oversight.
9	AI informs humans only if the system decides.	Passive observer unless the system signals issues.	Fully autonomous, decides when to inform a human.	Exception handling for critical decisions.	Extremely high: Trust in AI's judgment on when to involve humans.
10	Complete autonomy: AI operates independently of humans.	No involvement: addresses failures if needed.	Fully autonomous, acts without human input.	Pre-set system design and regulatory oversight.	Absolute: Trust AI to function independently and reliably.

evolution must be supported by thoughtful transparency, context-aware feedback, and clear steps for what happens when the system fails or needs human intervention. When working with developers or stakeholders, use this table to map:

- Who is **really making the decision** at each stage: Human, AI, or both?
- How much control should users have? When and how should they be informed?
- What kind of **trust signals** should be embedded in the UI (e.g., explanations, logs, override options)?

Ultimately, the goal isn't to eliminate complexity, but to **surface the right complexity at the right time**, supporting decision-making without overwhelming users. As designers, we must ask: are we helping users think less about what doesn't matter so they can focus on what does? Or are we simply burying complexity behind automation?

As AI systems grow more adaptive and less reliant on constant input, we must reframe the conversation around **agency**, **trust**, and **long-term impact**. Autonomy must be designed with care, for transparency, safety, and for the evolving relationship between humans and intelligent systems.

8.2.1 Trust Calibration: Balancing Expectations and Capabilities

Trust calibration means aligning user expectations with system performance, thereby avoiding the pitfalls of overtrust and distrust. Trust in AI centers around believing that relying on it won't leave us vulnerable in uncertain or high-risk situations. Trust calibration refers to the extent to which our trust in AI aligns with its actual capabilities. Users may place too much faith in AI when their trust surpasses the system's actual capabilities. Conversely, they may lose faith in the system if they feel uncertain about the AI's effectiveness.

Overtrust (Misuse)

Users place excessive reliance on the AI, assuming flawless performance. For instance, overtrust in autonomous vehicles can lead to accidents if users neglect to monitor the system during unexpected scenarios.

Distrust (Disuse)

Users underestimate or reject the system's capabilities due to a lack of understanding or transparency, resulting in missed opportunities. For example, healthcare professionals may distrust opaque diagnostic tools, opting instead for manual methods despite AI's potential advantages.

Both scenarios highlight the importance of fostering accurate mental models to ensure that users understand how the system operates and when to rely on it. So, how can designers help users calibrate their trust in a way that aligns with system performance?

8.2.2 Explainability and Control: The Foundations of Trust

This is where explainability and user control—two key principles emphasized by Kore—become essential. Clear communication regarding data use and transparent explanations of AI operations help users form accurate mental models. This clarity enables them to determine when to exercise caution or take manual action and how much they can rely on the system. Misalignment in trust, whether excessive or insufficient, can lead to annoyance, misuse, or even complete abandonment of the system. Additionally, feedback loops and personalization features further empower users, even within highly automated environments.

Explanations play a central role in bridging the gap between users' mental models and the actual workings of an AI system. Basic explanations are sufficient for low-automation systems, focusing on high-level overviews of functionality. However, as automation increases, the need for detailed explanations grows, encompassing decision-making processes, data sources, and potential risks. Figure 8.2 illustrates the interplay between automation levels and the depth of explanation required to align user expectations with system capabilities.

Anticipatory systems should strike a careful balance between automation and user control while ensuring transparency to prevent misuse or disuse. This canvas illustrates the effects of excessive trust or skepticism toward AI systems and formulates strategies for adjusting user trust. Such calibration is essential, as it influences the system's accountability, particularly when errors or failures occur.

To illustrate how trust calibration can be applied in practice, consider a Level 7–8 automated system designed to anonymize personal data from court decisions. The following example illustrates how the trust calibration canvas can help identify the

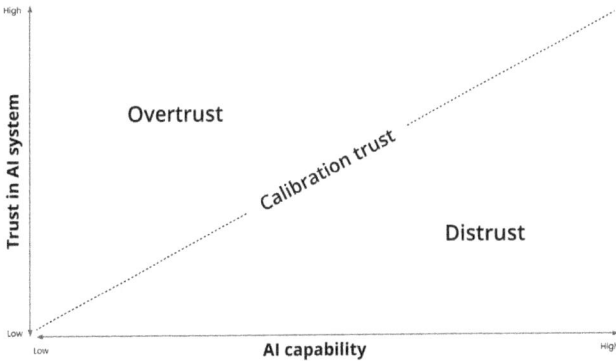

FIGURE 8.2   Trust calibration template according to Kore [37].

right balance between automation, user control, and transparency in such high-stakes applications.

This trust calibration process also aligns with evolving regulatory standards, such as the **ISO/IEC 42001:2023** checklist [41], which emphasizes transparency, integrated risk management, and human oversight throughout the AI lifecycle. Similarly, the **EU AI Act** (euaiact.com) highlights the importance of explainability and control, particularly in high-risk applications, and the Organisation for Economic Co-operation and Development **(OECD) AI Principles** [42] reinforce these values, advocating for responsible innovation that protects human rights and democratic values.

By providing timely, context-sensitive explanations, anticipatory systems can ensure users perceive AI as reliable collaborators rather than autonomous decision-makers. This new legislative framework mandates that AI systems used in critical sectors, such as healthcare, transportation, and finance, must provide clear and detailed explanations of their decision-making processes. Businesses must ensure that these systems are accurate, reliable, and transparent, with robust mechanisms for human oversight to prevent misuse and disuse.

Notably, the recent proposal passed by the EU Parliament on March 13, 2024, has significant implications for European providers of AI systems. The proposed EU AI Act classifies AI systems based on the potential risks and harms they may pose:

On the other hand, the OECD has established five value-based AI principles to promote the use of innovative yet trustworthy AI solutions that respect human rights and democratic values. They were adopted in May 2019 to set standards for AI that are practical and flexible enough to stand the test of time. The five principles are the following:

- **Principle 1.1: Inclusive growth, sustainable development, and well-being**: "Stakeholders should proactively engage in responsible stewardship of trustworthy AI in pursuit of beneficial outcomes for people and the planet, such as . . . reducing economic, social, gender and other inequalities."

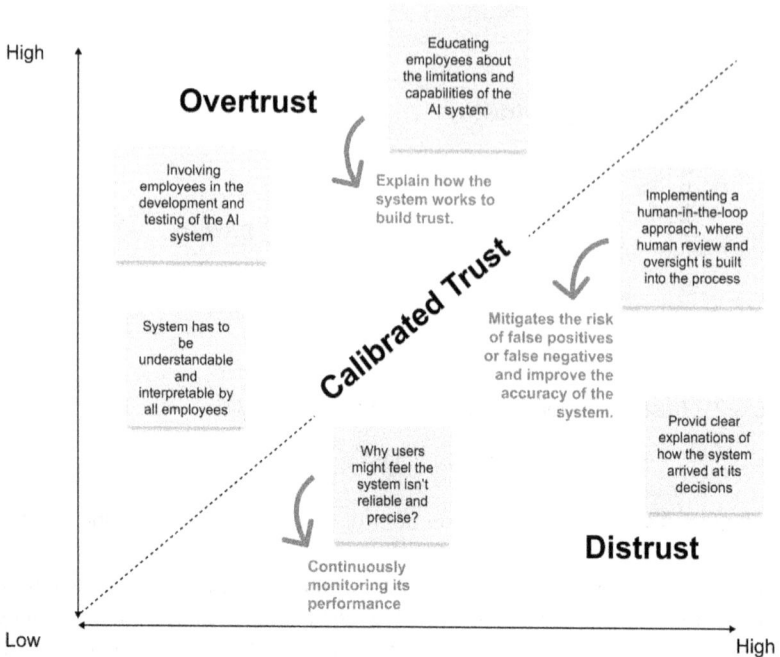

FIGURE 8.3   Illustration of trust calibration in an automated court administration decision system, demonstrating how to balance automation, user control, and transparency.

- **Principle 1.2: Human-centered values and fairness**: "AI actors should respect the rule of law, human rights and democratic values, throughout the AI system lifecycle. . . . To this end, AI actors should implement mechanisms and safeguards, such as the capacity for human determination, that are appropriate to the context and consistent with the state of art."
- **Principle 1.3: Transparency and explainability**: "to foster a general understanding of AI systems; to make stakeholders aware of their interactions with AI systems."
- **Principle 1.4: Robustness, security, and safety**: "AI actors should ensure traceability, including in relation to . . . decisions made during the AI system lifecycle."
- **Principle 1.5: Accountability**: "AI actors should be accountable for the proper functioning of AI systems and for the respect of the above principles, based on their roles, the context, and consistent with the state of art."

In summary, the OECD AI Principles provide a robust ethical foundation grounded in transparency, accountability, and human-centered values, aligning closely with the objectives of anticipatory design. These principles and emerging regulations, such as the ISO/IEC 42001:2023 checklist and the EU AI Act, emphasize the importance of transparency, human oversight, and explainability, especially in high-risk contexts.

TABLE 8.3 Risk-Based Classification of AI Systems Under the Proposed EU AI Act, Including Examples and Corresponding Regulatory Obligations.

RISK LEVEL	EXAMPLES	OBLIGATIONS
Unacceptable risk	— Social scoring by governments — Exploitation of vulnerable groups — Subliminal manipulation techniques	Prohibited outright
High risk	— AI in critical infrastructure (e.g., energy, transportation) — Medical devices — Hiring algorithms — Law enforcement (e.g., facial recognition) — Migration control	Must meet strict requirements: • Risk management • High-quality datasets • Record-keeping • User transparency • Human oversight • Security and accuracy
Limited risk	— Chatbots — AI that generates or manipulates content (e.g., deepfakes)	Must inform users that they are interacting with AI
Minimal risk	— AI in video games — Spam filters	No legal requirements, but voluntary codes of conduct encouraged

Collectively, they aim to ensure that AI systems remain safe, fair, and aligned with user needs. As anticipatory systems become more autonomous, trust calibration will be essential, achieved through tailored explanations, meaningful user control, and precise performance metrics that help users assess system reliability. These frameworks protect users and guide designers in creating AI experiences that empower rather than alienate.

TL;DR

As AI systems become more autonomous, designing for trust becomes more complex and essential. This chapter examines how anticipatory systems can preserve user agency by striking a balance between automation, explainability, transparency, and user control.

- **Automation ≠ delegation**: Over-automating tasks without offering insight or flexibility can disempower users and break trust.
- **Trust must be calibrated**: Designers must align system behavior with user expectations through clear mental models, contextual explanations, and override mechanisms.
- **Augmentation vs. automation**: These models must be balanced, not just technically but behaviorally, based on the task's complexity and sensitivity.

- **Framework extension**: Building on Kaushik and Kore's models, the chapter introduces a matrix that maps the evolution of roles, responsibilities, and trust with automation.
- **Explanation depth matters**: The higher the automation, the more thorough and transparent the system's explanation must be.
- **Legal and ethical grounding**: Standards like the EU AI Act, ISO/IEC 42001:2023, and the OECD AI Principles provide guardrails that support human oversight, ethical design, and accountability.

The Science of Behavior Change

9

So far, it's clear that although predicting user needs sounds appealing, it presumes a predictability in human behavior that often isn't realistic, ultimately undermining the personalization that businesses seek. The previous chapter examined how autonomy must be balanced with transparency and control to foster trust and sustain user engagement over time. To achieve this, the chapter presents behavioral science as an essential tool for creating effective anticipatory experiences. It offers frameworks and strategies to comprehend, forecast, and enhance user behavior through goal-focused design that aligns system interactions with users' objectives and values.

Designing anticipatory experiences requires more than accurate data predictions—it demands a deep understanding of human behavior. While the previous chapter outlined the promise and pitfalls of anticipatory design, it also highlighted a critical challenge: human behavior rarely aligns neatly with statistical models alone. People's actions, motivations, and decisions are dynamic, influenced by context, emotions, and changing needs. Behavioral science provides designers with vital insights into these nuances, ensuring that anticipatory systems resonate meaningfully with users.

Behavioral science sits at the intersection of psychology, technology, and user experience, enabling designers to encourage healthier habits, facilitate better decisions, and support lasting changes. Several influential psychological frameworks are essential to achieve this: Fogg's **behavior model (B=MAP)**, Prochaska's **transtheoretical model (TTM)**, and Thaler and Sunstein's **Nudge theory**. These frameworks provide a comprehensive toolkit for designers to create anticipatory experiences that align with human behavior and systemic contexts, rather than struggling against them.

DOI: 10.1201/9781003642800-11

9.1 A FRAMEWORK FOR UNDERSTANDING BEHAVIOR: FOGG'S B=MAP MODEL

Behavioral scientist BJ Fogg introduced a deceptively simple yet powerful formula in 2009 to explain human behavior [43].

Behavior (B) = Motivation (M), Ability (A), Prompt (P)

Behavior happens when motivation, ability, and a prompt come together simultaneously. When a behavior does not occur, at least one of those three elements is missing [42]. According to Fogg, Behavior (B) is the result of three interacting elements:

B=MAP

- **M: Motivation** → How strongly users desire to act.
- **A: Ability** → How easy or difficult it is for users to act.
- **P: Prompt** → The trigger or cue that initiates the action.

This model emphasizes the importance of thoroughly considering all three elements when designing interventions to promote behavior change. Fogg argues that all three elements must be present simultaneously for a behavior to occur. This dynamic interplay underscores the designer's role in balancing these elements. If any are missing, behavior is unlikely to happen. For example:

- **High motivation and ability, but no prompt** → No behavior. The user is unsure when or how to act.
- **Strong prompt and high motivation, but low ability** → Behavior is unlikely to happen due to difficulty.
- **Clear prompt and high ability, but low motivation** → Behavior fails due to a lack of interest. The user lacks the drive to engage in the behavior.

Fogg's formula is not just a tool for explaining behavior but a guide for crafting interventions that drive meaningful change. Each component of the model plays a distinct role in shaping behavior.

9.1.1 Motivation

Motivation represents the degree to which an individual is driven to act and drives user behavior along three critical dimensions:

- **Pleasure versus pain**: These are immediate emotional drivers that prompt an individual toward or away from a behavior.

- **Hope vs. fear**: These are anticipatory emotions tied to potential future outcomes. Hope motivates action by promising a positive result, while fear compels action to avoid negative consequences.
- **Social acceptance versus rejection**: The need for social belonging plays a critical role, as individuals are often motivated by the desire to align with societal norms or avoid ostracism.

By leveraging these dimensions, designers can create interventions that connect with users on an emotional and social level, thereby enhancing their motivation to take action.

9.1.2 Ability

Ability involves simplifying user actions. Fogg emphasizes simplicity as a critical factor in increasing ability, identifying six elements that can either enable or hinder an action:

- **Time**: How much time is required to perform the behavior?
- **Money**: The financial cost associated with the behavior.
- **Physical effort**: The level of exertion needed.
- **Mental effort**: The cognitive load required.
- **Social deviance**: How aligned is the behavior with social norms?
- **Non-routine**: Whether the action is familiar or part of the individual's habits.

To improve *ability*, designers need to dismantle barriers linked to these elements. Headspace, a widely used meditation and mindfulness app, exemplifies these principles well. It recognizes that user convenience is crucial for sustaining lasting behavioral change, intentionally reducing obstacles to meditation practice.

The app streamlines meditation through short, structured, and guided sessions, significantly reducing **time** commitments. It provides free introductory content and affordable plans, minimizing **financial** constraints. Its guided audio sessions require minimal **physical effort**, allowing users to practice comfortably in diverse environments.

Moreover, Headspace employs user-friendly interfaces and straightforward instructions, carefully minimizing **mental effort** and complexity often associated with meditation. By framing meditation as mainstream and beneficial, it also reduces **social deviance**, aligning mindfulness practices with widely accepted norms. Additionally, these accessible sessions provide seamless support for habit-building, addressing the **non-routine** nature of meditation for newcomers.

Together, these intentional design choices significantly enhance users' capacity to consistently engage in mindfulness, fostering long-term habit development and effective behavioral change.

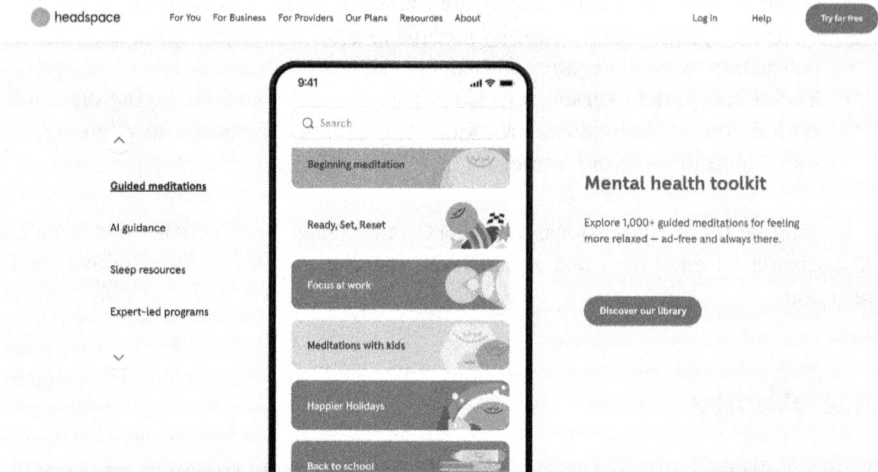

FIGURE 9.1　A 2025 screenshot of the Headspace App website.

Source: headspace.com

9.1.3 Prompts

Prompts are the triggers or cues that initiate behavior. Without a prompt, even highly motivated individuals with the ability to act may not perform the desired behavior. Fogg categorizes prompts into three types:

- **Sparks**: These are emotional triggers designed to boost motivation and drive. For example, a health app might use a motivational message to encourage users to meet their daily step goal.
- **Facilitators**: Prompts make a behavior easier by addressing barriers to ability. A guided tutorial in a fitness app can simplify complex exercises, thereby enhancing the user's ability to perform them effectively.
- **Signals**: These are simple reminders or cues, such as notifications, that prompt users to take action.

Effective prompts match motivation and ability appropriately. Overreliance on high motivation without considering ability can backfire if users feel confused or overwhelmed.

The B=MAP model provides a scientifically grounded and actionable framework for designing behavior change interventions. As Fogg states, behavior is not merely a result of intrinsic willpower but a product of well-designed systems that align motivation, simplify ability, and deliver timely prompts. By focusing on the interplay between motivation, ability, and prompts, designers can create systems that effectively guide users toward desired behaviors.

9.2 INTEGRATING ADDITIONAL FRAMEWORKS

Fogg's model lays a solid foundation; however, understanding human behavior, particularly in AI contexts, requires additional frameworks. Two frameworks that enhance behavioral insights are Prochaska's **TTM**, also known as the Stages of Change, and **Nudge theory**. The TTM illustrates an individual's readiness to change, while Nudge theory offers an innovative approach to influencing behavior through subtle interventions that encourage better choices without coercion. These frameworks deserve exploration, as AI enhances behavior change by providing features such as real-time feedback, personalization, and automated nudges.

9.2.1 Prochaska's Transtheoretical Model

Complementing B=MAP with TTM provides a stage-based approach to understanding behavior change. It recognizes behavior change as a process that unfolds over time, involving progress through several stages. TTM conceptualizes behavior change as a process involving progression through five stages:

1. **Pre-contemplation**: Not ready—unaware or unmotivated
2. **Contemplation**: Getting ready—recognizing the need to change
3. **Preparation**: Ready—planning and setting intentions
4. **Action**: Making changes—actively engaging in new behaviors
5. **Maintenance**: Keeping up with changes—sustaining long-term changes

A notable example of an AI anticipatory experience that struggled to address the full spectrum of user stages is the case of **Noom**, a digital health app designed to support weight management and healthy living.

FIGURE 9.2 Representation of the Noon App service for weight loss [44].

Noom's AI-driven platform is effective for users already in the **Action** stage of behavior change, offering structured coaching, habit tracking, and daily lessons. However, it has faced challenges in engaging individuals in earlier stages, such as **pre-contemplation** or **contemplation**, where initial motivation and readiness to change must still be nurtured. Many new users encounter prescriptive recommendations and rigid behavior targets, such as daily weigh-ins or calorie goals, which can feel over-whelming or misaligned with their mindset. This can discourage those who are not yet fully committed to change, especially when the app's branding as a "non-diet" solution doesn't align with its structured tracking expectations.

Moreover, Noom's strategies have also shown limitations in supporting **long-term behavior maintenance**. Sustained success depended heavily on forming healthy habits, which Noom's coaching model did not always effectively reinforce. Users who achieved short-term goals often reported **a lack of adaptive support** to accommodate changes in motivation, setbacks, or shifting lifestyle contexts. Rather than evolving with users' progress, the system continued to deliver generalized advice that didn't reflect individual fluctuations. As a result, users often disengaged after reaching initial milestones, citing limited support for relapse prevention or long-term habit reinforcement.

9.2.1.1 Beyond Action: Supporting All Stages of Change

This example illustrates the need for anticipatory systems to design **stage-sensitive interventions** that evolve over time. For effective behavior change, systems like Noom must support users before, during, and after change occurs—offering low-pressure exploration for newcomers, dynamic goal adjustment for experienced users, and sustained, personalized feedback for long-term success. This situation highlights the crucial need for behavior change systems to cater to users at *all* stages of the TTM—from pre-contemplation to maintenance. Interventions must be tailored to each stage: offering curiosity-driven prompts and reflective nudges for early stages, and delivering responsive coaching and progress reinforcement during and after behavioral shifts.

Notably, this challenge is not unique to Noom. Competing services, such as **Yazio**, have faced similar critiques, focusing heavily on structured tracking and short-term goal reinforcement while neglecting the deeper psychological journey of change. These platforms often rely on prescriptive routines and behavior prompts that may be ineffec-tive for already motivated users. However, they struggle to support those who are just beginning to consider change or trying to sustain it in the long term.

These examples reveal a broader trend in digital health platforms: while technically reducing cognitive effort, they are often behaviorally misaligned. They optimize for action but fail to design for the *journey*—the shifting motivations, cognitive readiness, and emotional context that shape behavior over time. Without stage-sensitive, adaptive interventions, even the most advanced systems risk disengagement and abandonment. True anticipatory design must be grounded not only in prediction but also in empathy for where users are in their behavior change journey and where they need help next.

9.2.2 Powering Up Frameworks: BMAP Plus TTM

The shortcomings of Noom and Yazio highlight what can go wrong when AI-driven systems fail to adapt across the full spectrum of behavior change stages. In contrast, some platforms offer promising insights into what becomes possible when behavioral frameworks are applied more holistically. One such example is **Smart Sparrow,** an adaptive learning platform that appears to integrate both Fogg's B=MAP and the TTM to support users at various stages of change.

Where Noom and Yazio relied on rigid interventions aligned primarily with users already in the action phase, Smart Sparrow adopts a more flexible, stage-sensitive approach—proactively engaging users before, during, and after behavioral shifts.

For instance, reducing the mental effort required to complete a task, such as through intuitive interfaces or guided instructions, can significantly increase the likelihood of behavior change. For instance, Smart Sparrow (smartsparrow.com) provides educators with tools to create personalized learning experiences for their students. This service illustrates the application of behavioral science principles by supporting users across various stages of the TTM. In the pre-contemplation and contemplation stages, Smart Sparrow uses intuitive onboarding processes and emotionally engaging content to raise awareness and spark curiosity among educators and learners who may not yet recognize the need for adaptive learning. For example, US-based users may encounter the Dot Resiliency Series, a social–emotional learning initiative that encourages reflection through emotionally resonant simulations. More broadly, Smart Sparrow showcases performance metrics and learning patterns that prompt consideration of new teaching strategies, helping users identify the potential value of personalized tools.

As users move into the preparation and action stages, Smart Sparrow provides adaptive courseware and timely prompts that align with Fogg's B=MAP principle of delivering behavior-supporting triggers at the right time. Educators are guided by template-driven tools and notifications, such as alerts about increased student engagement when interactive simulations are added, which reduce cognitive and technical friction while sustaining motivation.

Smart Sparrow continues to adapt in the maintenance stage by leveraging learning analytics and dynamic feedback loops. The platform enables ongoing tracking of student progress, offering targeted recommendations that help educators adjust content and support individual learning needs over time. This responsiveness ensures that both motivation and relevance are sustained, illustrating how anticipatory learning systems can foster meaningful, long-term engagement through personalization and behavioral alignment.

9.2.3 Nudge Theory: Subtle Guidance Without Coercion

Frameworks such as TTM and B=MAP provide essential principles for behavior change, but their implementation in anticipatory systems requires a deeper understanding of user transitions across stages over time.

Developed by Thaler and Sunstein in 2008 [45], the Nudge theory complements the previous frameworks by advocating for subtle, timely interventions that guide individuals toward better choices without imposing restrictions. Therefore, it emphasizes noncoercive, timely interventions that improve decision quality by:

- Aligning incentives with desired behaviors.
- Providing immediate feedback to reinforce actions.
- Simplifying choices for complex decisions.
- Making goals and progress visible through feedback loops.

These elements turn nudges into powerful tools that support behavior change and sustain motivation. Personalized nudges, grounded in data and tailored to the user's stage of change, enhance the likelihood of sustained engagement and practical learning.

For instance, analyzing Noom and Yazio reveals that ignoring users in the pre-contemplation and contemplation stages leads to disengagement and lost chances. To fully leverage TTM's capabilities, anticipatory systems must be attuned to users' evolving mental states and stages of change, incorporating advanced features like real-time feedback, PA, and personalized experiences. That's why integrating Nudge theory is crucial for anticipatory systems, offering a necessary layer of subtle persuasion without coercion, complemented by real-time decision support that stage-based models and motivational frameworks alone cannot adequately provide.

9.3 DESIGNING BEHAVIOR CHANGE WITH AI

To create impactful anticipatory experiences, we require a more nuanced understanding of anticipatory systems. For this reason, I suggested incorporating B=MAP, TTM, and Nudge theory behavioral models to fill gaps such as initial engagement, long-term adaptability, and transparency. These serve as guiding principles for effective behavior change design:

- **Dynamic adaptability**: Design systems that evolve with user behaviors, incorporating feedback loops and recalibration.
- **Stage-specific engagement**: Address pre-contemplation through exploratory prompts and maintenance through personalized, evolving strategies.
- **Transparency-driven trust**: Foster user confidence by making predictions explainable and actionable.

Designers can create AI-driven solutions that are not just effective but also empathetic to human behavioral complexities by utilizing these frameworks and principles. Upon reflecting on these behavioral frameworks, we can infer that three fundamental principles of behavior change applicable to design are user agency, personalization, and timely interventions.

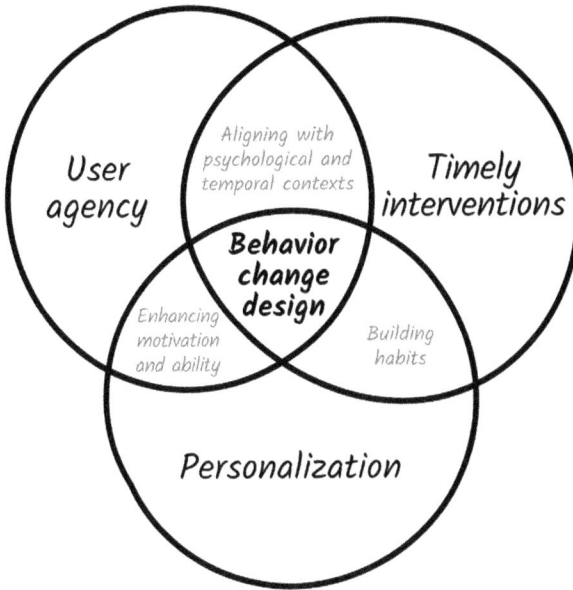

FIGURE 9.3 Incorporating three behavioral science frameworks into designing anticipatory experiences.

If we combine the previous frameworks, we now have an effective way to design successful anticipatory experiences that sustain behavior change:

- We need **user agency** to emphasize control and autonomy, thereby fostering trust.
- We need **personalization** to adapt to users' stages of change, aligning with their specific needs.
- Lastly, we need **timely interventions** to guide users effectively and to guide decisions at the right moment.

When these three models overlap, we create systems that empower users, adapt to their needs, and guide them toward long-term results.

9.3.1 User Agency in Anticipatory Design

The first principle—**User Agency**—draws upon Fogg's behavior model, which suggests that behavior is a function of three factors: motivation, ability, and triggers. A critical takeaway from Fogg's research is that users need to believe they have control over their actions. When users perceive a system as coercive or manipulative, they are less likely to engage effectively with it. Empowering users to make decisions maintains their autonomy while guiding them toward better choices. Behavior change is most

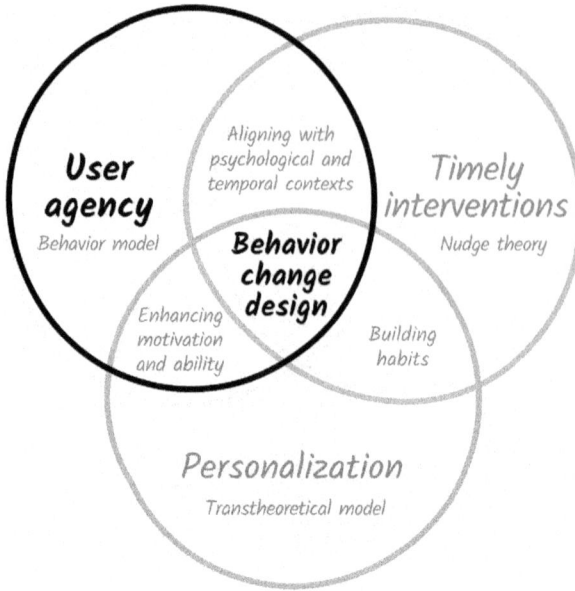

FIGURE 9.4   Highlight of user agency in the behavior science Venn diagram.

successful when individuals feel in command, a factor crucial for nurturing long-term engagement and ensuring trust in the system.

In anticipatory design, systems are designed to predict and preemptively respond to user needs. Nevertheless, user agency is essential to prevent overpromising and underdelivering and to mitigate the risk of losing control.

Use the following checklist to ensure that your anticipatory system empowers users, respects autonomy, and builds trust through transparency and control.

1. **Give Users Control Over Predictions and Actions**
   - ☐ Allow users to override automated predictions or actions.
   - ☐ Offer flexible controls for how predictions are generated or acted upon.
   - ☐ Let users set preferences that influence future system behavior.

2. **Provide Options for Customization and Transparency**
   - ☐ Clearly explain why specific predictions or suggestions are made (e.g., based on usage patterns or behavioral data).
   - ☐ Offer users the ability to adjust or refine the system's logic or recommendation parameters.
   - ☐ Make the system's reasoning and data sources visible and understandable.
   - ☐ Disclose when the system's prediction confidence is low (especially when the user falls outside the training data distribution. Explain this in user-friendly, verbal, or visual terms, and offer options to further adjust, ignore, or personalize the suggestion).

3. Avoid Overreliance on AI Decisions
- ☐ Ensure that users can always reject or modify system recommendations.
- ☐ Avoid default behaviors that act without user input in high-stakes or personal contexts.
- ☐ Regularly test whether your predictions are helpful or overbearing to different user segments.

4. Provide Active Feedback on Predictions and Results
- ☐ Include interactive feedback loops (e.g., "Did this automation support you?").
- ☐ Adapt future suggestions based on user feedback or adjustments.
- ☐ Demonstrate to users how their preferences or actions have influenced the system's behavior and evolution.

9.3.2 Personalization in Anticipatory Design

The second principle—**Personalization**—aligns with Prochaska's TTM, which emphasizes that behavior is not static—individuals progress through different stages when adopting new behaviors. Personalized interventions that align with a user's specific stage of change are significantly more effective than generic, one-size-fits-all approaches.

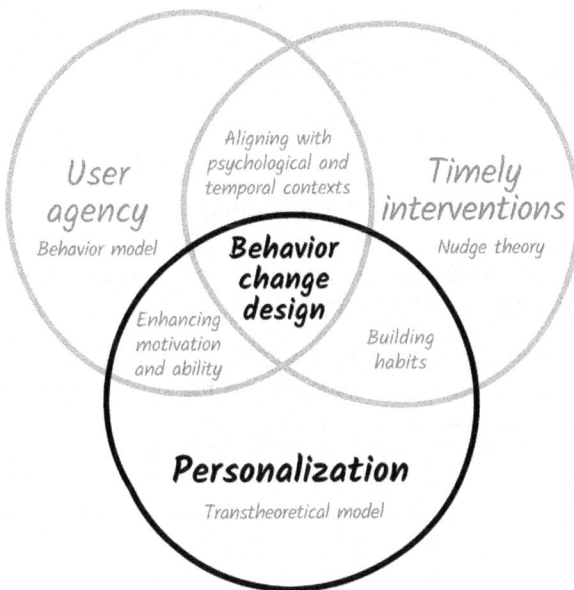

FIGURE 9.5   Highlight of personalization in the behavioral science Venn diagram.

Personalized interventions increase the likelihood of behavior change by aligning with the user's current needs and goals. Personalization lies at the heart of anticipatory design, as it adapts interventions to align with each user's unique needs and context. By accounting for individual preferences, personalization ensures that suggestions remain both relevant and actionable. Without this tailored approach, systems risk appearing generic and intrusive, ultimately diminishing their effectiveness and user engagement. For instance, a learning app that dynamically adjusts its difficulty level based on the user's proficiency delivers a personalized experience that maximizes engagement and long-term success.

Use the following checklist to ensure that your anticipatory system delivers relevant, adaptive, and user-aligned experiences by tailoring interventions to individual needs, preferences, and behavioral stages.

1. **Align Predictions With User Context**
   - ☐ Use behavioral stage models (e.g., TTM) to tailor predictions to an individual's readiness.
   - ☐ Analyze recent behaviors or context to refine what and when to suggest.
   - ☐ Adapt content to individual goals, routines, and constraints.

2. **Support Gradual Learning and System Influence**
   - ☐ Introduce features and interventions progressively.
   - ☐ Let users scale the depth or frequency of system interventions over time.
   - ☐ Provide onboarding and light-touch support for new behaviors.

3. **Avoid One-Size-Fits-All Frustration**
   - ☐ Personalize recommendations based on user preferences and limitations (e.g., dietary restrictions, schedule, accessibility needs).
   - ☐ Avoid generic or repetitive nudges that feel irrelevant or misaligned with user goals.
   - ☐ Use inclusive language and content tailored to different learning styles or goals.

4. **Allow for Adjustments and Customization**
   - ☐ Let users adjust the type, timing, and frequency of nudges or interventions.
   - ☐ Provide easy-to-access controls to modify personalization preferences.
   - ☐ Offer transparency about how and why content or actions are personalized.

5. **Update Personalization Dynamically**
   - ☐ Continuously update recommendations based on evolving behaviors and engagement patterns.
   - ☐ Detect and adapt to changes in user context, such as shifting goals or new constraints, to ensure optimal performance.
   - ☐ Build in feedback loops that let users refine or confirm system assumptions.

9.3.3 Timely Interventions in Anticipatory Design

Finally, when designing anticipatory experiences, it is essential to consider the principle of **timely interventions**. Drawing on Thaler and Sunstein's Nudge theory, this concept

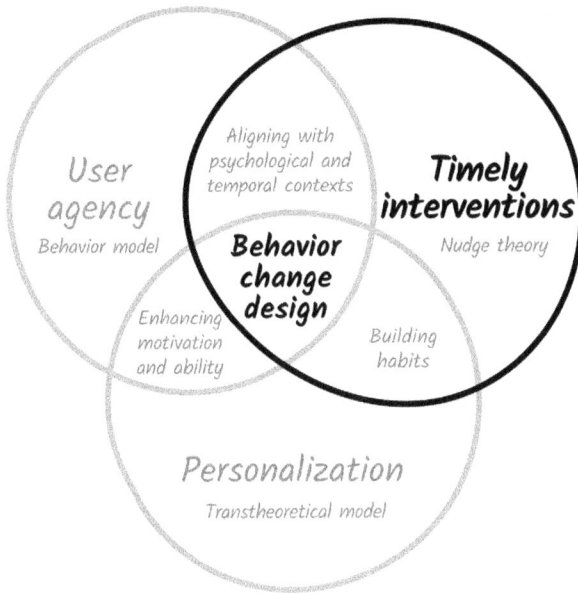

FIGURE 9.6 Highlight of timely interventions in the behavior science Venn diagram.

emphasizes delivering interventions at precisely the right moment to encourage positive behavior without coercion. Timely nudges significantly enhance the likelihood of achieving desired behaviors by targeting moments when users are most receptive. For instance, sending well-timed reminders to save money or take the stairs can subtly guide individuals toward healthier or more beneficial choices, making the experience both seamless and effective.

Nudging is especially important for anticipatory design because timely, context-aware interventions are at the heart of what makes anticipatory systems effective and user-centric. Anticipatory design relies on predicting user needs and acting preemptively, but these predictions can feel intrusive or irrelevant without well-timed and subtle nudges.

Use the following checklist to ensure that your anticipatory system delivers timely, respectful, and effective nudges that transform predictions into meaningful actions without overwhelming or disempowering users. Each section begins with a principle and is followed by specific design actions you can implement.

1. **Bridge the Gap Between Prediction and Action**
   - ☐ Clearly connect predictions to tangible user benefits (e.g., "Lowering your thermostat saves $15 this month").
   - ☐ Use prompts to translate insights into timely, actionable suggestions.

2. **Ensure That Nudges Are Contextually Relevant**
   - ☐ Confirm predictions with users before acting on inferred patterns.

☐ Avoid making suggestions that are irrelevant or outdated, as they may feel invasive or annoying.

☐ Explain the rationale behind suggestions to maintain relevance and build confidence.

3. Allow Personalization and Adjustment

☐ Enable users to customize nudge timing, frequency, and format (e.g., silent vs. pop-up reminders).

☐ Provide intuitive settings to adjust the sensitivity and relevance of prompts.

☐ Enable users to control the types of nudges they receive, such as reminders versus motivational cues.

4. Respect Autonomy While Offering Guidance

☐ Frame nudges as optional suggestions, not enforced actions or defaults (e.g., "Would you like me to").

☐ Always provide a straightforward way to decline, skip, or modify the proposed action.

5. Use Nudges to Mitigate Over-Prediction

☐ Use nudges as soft checkpoints when system predictions might be wrong.

☐ Design fallback responses for edge cases where prediction might fail.

☐ Disclose uncertainty where appropriate (e.g., "You might need" instead of definitive statements).

6. Reduce Cognitive Load Without Over-Automating

☐ Simplify decisions through focused, low-effort prompts (e.g., "Tap to save $50 this month").

☐ Present clear, low-friction calls to action that avoid overwhelming users with options.

Nudges are essential in anticipatory design because they bridge the gap between the system's predictions and the user's actions. This ensures that interventions are timely, contextually relevant, and respectful of user autonomy. By effectively leveraging nudges, anticipatory systems can guide users toward better outcomes without being overbearing, intrusive, or diminishing trust. This aligns perfectly with the principles of behavior change design—user agency, personalization, and timely interventions—while making anticipatory design more intuitive and human-centered.

9.4 CRITIQUES AND OPPORTUNITIES IN BEHAVIOR CHANGE DESIGN

While frameworks like Fogg's behavior model and Prochaska's TTM model offer robust foundations for understanding behavioral change, applying them in anticipatory systems reveals common gaps, particularly in sustaining user engagement across different stages of change and during habit formation.

Do you recall Vi Sense from Chapter 7? This case is especially instructive because it highlights what can go wrong when anticipatory systems fail to align with users' psychological readiness for change. Although Vi Sense succeeded in attracting users already in the *Action* stage of the TTM—those already motivated to run outdoors— it overlooked users in earlier stages, such as *Pre-contemplation* and *Contemplation*, where motivation must first be sparked and gradually nurtured. The system successfully attracted users who were already motivated to engage in fitness activities—those in the action stage of Prochaska's TTM. Without low-pressure, curiosity-driven prompts or exploratory tools to make users consider their fitness goals, many potential users disengage early in the journey.

Vi Sense also struggled with *habit formation*. Research shows that establishing a new behavior takes at least 21 days of consistent engagement [46]. Yet the system lacked a structured behavioral plan or feedback loop to support this formative period. Instead of offering adaptive, context-aware nudges—based on routines, real-time weather, recent performance, or health condition—the coaching felt static and misaligned with user needs. For users recovering from setbacks or progressing faster than expected, the generic feedback led to a loss of relevance and, eventually, disengagement.

From the perspective of Fogg's B=MAP model, Vi Sense also fell short. It provided motivational audio feedback but failed to reduce friction around *ability* or offer timely, varied *prompts*. For example, if a user missed a run due to rain or sleep disruption, the system didn't adjust by recommending a short indoor session or walking alternative. Nor did it acknowledge real-world obstacles, which are central to sustaining behavior change.

Vi's shortcomings also highlight the importance of *transparency* in building trust. Users were rarely told why the system made specific recommendations or how actions contributed to long-term goals. For anticipatory systems, transparent rationale is essential. A simple message, such as "Running three times this week helps build your 21-day habit streak," could have reinforced users' sense of progress and deepened their trust in the system's guidance.

In essence, despite its innovative features, the solution felt rigid and overly tailored to beginner runners. Vi failed to anticipate user performance or adapt to changing conditions, thereby undermining its promise of personalized, adaptive coaching. This created a gap between the vision of truly responsive, real-time AI coaching and the reality of its limited and inflexible functionality. As a result, many users became frustrated and ultimately disengaged, contributing to the broader decline in adoption and ultimately leading to the product's shutdown.

Ultimately, Vi Sense illustrates a missed opportunity: the system failed not because of inadequate technology but because the behavior change journey was poorly supported. To address this, anticipatory systems must embed behavioral science principles more effectively by:

• Designing stage-specific experiences to engage users across the TTM spectrum.

- Providing adaptive, context-aware nudges during the 21-day habit-forming window.
- Ensuring that motivation, ability, and prompt are in constant balance.
- Offering transparency and feedback to help users understand their progress.

By weaving together insights from B=MAP, TTM, and habit-formation literature, anticipatory systems can transform rigid predictions into dynamic, personalized guidance—empowering users not just to start but also to sustain meaningful behavior change. But let's be clear: designing for behavior change isn't about nudging users toward business outcomes—it's about helping them reach their own goals. When grounded in psychological insight, anticipatory systems become adaptive, empathetic allies in everyday decision-making.

TL;DR

Anticipatory systems often assume human behavior is predictable, but real behavior is complex, contextual, and fluid. This chapter introduces behavioral science as a foundational lens for designing AI systems that truly support change, grounding anticipatory design in how people actually think, feel, and act.

It explores three key frameworks:

- **Fogg's B=MAP model** emphasizes that behavior arises from motivation, ability, and timely prompts. It shows designers how to reduce friction and trigger action at the right moment.
- **Prochaska's TTM** offers a stage-based understanding of readiness to change, reminding us that behavior change is a journey that requires different supports at various stages.
- **Nudge theory** adds nuance by advocating for subtle, noncoercive interventions that guide decisions without overwhelming users.

The chapter argues that **user agency, personalization, and timely interventions** are essential design principles for sustained engagement. Case studies like Noom and Vi Sense demonstrate how even innovative AI systems can fail when they overlook behavioral science, including habit formation, stage-specific needs, and adaptive support.

Designers are encouraged to think beyond predictions and to support **meaningful, self-directed behavior change** through empathetic, adaptive systems that guide without control.

PART III

Anticipatory Design in Action

Approaches, Frameworks, and Practical Examples

From Theory to Method

10

Interpreting Anticipatory Design

So far, we've examined the idea of anticipation from various perspectives—its evolutionary background, psychological aspects, and its tumultuous introduction to the digital realm. We've observed how positive intentions can result in unintended consequences and why modern systems frequently fail to anticipate human needs effectively. In this chapter, I will start the transition from contemplation to implementation. What is necessary to create anticipatory systems that not only forecast but also care? That not only streamlines but also empowers? Let's delve into the methods, frameworks, and design practices that can realize this vision.

10.1 DIFFERENT COMPREHENSIONS OF ANTICIPATORY DESIGN

Biology and psychology inspire systems that learn, adapt, and evolve. Their insights offer rich metaphors here:

- **Biology** shows us how organisms survive by preparing for future threats and adjusting behaviors based on shifting environments.
- **Psychology** reveals how our brains constantly form expectations and refine them through feedback, helping us navigate a complex world.

Together, these perspectives guide us in designing systems that reduce mental load, support user intent, and behave like helpful companions, not controlling overlords. Let's explore how various thought leaders and frameworks interpret anticipatory design—and how they differ in their priorities, methods, and outcomes.

Celi and Colombi see anticipatory design as rooted in foresight and future thinking, emphasizing speculative exploration and cultural evolution [47]. Their approach includes the following:

- **Thinking in many futures**: Instead of trying to predict a single outcome, this approach explores *multiple possible futures.* It uses creative, exploratory methods to help designers navigate uncertainty and imagine how different scenarios could unfold.
- **Spotting trends as signals**: Trends are treated as "future prompts." Designers use them to make sense of cultural shifts, social signals, and early signs of change, turning vague patterns into meaningful design directions.
- **Metadesign lens**: The authors frame anticipatory design through a *metadesign* lens. That means embracing open-ended pathways rather than fixed outcomes, designing frameworks that evolve over time, not one-off solutions.
- **Layered exploration of the future**: Their method digs beneath the surface, beyond obvious trends, into deeper cultural beliefs, societal myths, and values. This multilayered thinking helps address messy, systemic problems (wicked problems) with more nuance.
- **Shaping culture through design**: They also emphasize how design can redefine meaning and influence culture. It's not just about solving problems—it's about shifting the stories, values, and symbols that shape how people imagine the future.

While this method fosters creative thinking, it tends to be quite abstract. It is immensely valuable for examining broad societal changes but can be challenging to apply directly to product design or user flows.

Cascini approaches anticipatory design through the ***Laws of Engineering Systems Evolution***, making it ideal for technical innovation and process optimization [48]. His approach focuses on the following:

- **Modeling the system clearly**: Begins by developing comprehensive models of a system's operations, allowing teams to grasp all its components and their interactions.
- **Identifying areas for improvement**: Focuses on recognizing the aspects of a system that can still be enhanced while avoiding unnecessary attention on elements that are functioning effectively or have already matured.
- **Designing with future change in mind**: Instead of only addressing current needs, this perspective anticipates future developments, guiding teams to make strategic design decisions by considering how products and systems may evolve over time.

Nonetheless, it overlooks the importance of human intent, changes in behavior, and the emotional subtleties of user experiences. I believe these are vital traits that anticipatory design should incorporate to blend human-centered methodologies with predictive technologies, thereby crafting solutions that resonate closely with user aspirations and goals.

Last but not least, **Morrison** and colleagues bring a critical design perspective that emphasizes the following [49]:

- **Rethinking relationships**: Exploring how people, technology, and the environment are connected—and how we can design for long-term, sustainable impact.
- **Designing across scales**: Tackling big challenges (like climate change) by thinking across different levels—from individual interactions to global systems.
- **Designing for diverse futures**: Understanding that people imagine and experience the future differently based on their culture, background, and context.
- **Exploring "What-Ifs"**: Using storytelling and imaginative design (like design fiction) to challenge assumptions, provoke thought, and explore possible futures.

This framework is thought-provoking and essential for large-scale systemic change. Still, like Celi and Colombi, it tends to remain theoretical and less actionable for day-to-day UX or product teams. Morrison's innovative and reflective approach addresses power dynamics, sustainability, and participatory politics. However, it prioritizes speculative and critical design over actionable, practical systems that deliver immediate value to users. While their macro-perspective is valuable for broader societal and cultural considerations, our understanding focuses on a more micro evaluation that creates simultaneously predictive and user-aligned systems, emphasizing trust, transparency, and adaptability.

10.2 ANTICIPATORY DESIGN AS A DESIGN METHODOLOGY

Anticipatory design is the process that enables the creation of anticipatory experiences. However, from a design perspective, its methodology is still in active development. While the ambition is clear—designing systems that can think and act ahead—the paths and methods to achieving this are still emerging. Scholarly discussions around anticipatory design remain fragmented, and more comprehensive, practice-oriented exploration is needed to understand its capabilities and limitations fully.

Designers and non-designers working in this space must consider a wide range of interconnected factors, including technological, social, cultural, and environmental. Anticipatory systems don't exist in isolation; they operate within real-world ecosystems that shape, constrain, and evolve with them. To navigate this complexity, designers must work across three key dimensions: the **technological**, the **people-social**, and the **environmental** [50]. Ignoring any of these dimensions risks creating technically functional solutions that are socially disconnected or environmentally unsustainable. The designer's role, then, is to serve as a bridge—linking innovation to real, lived contexts in useful, ethical, and resilient ways.

As with the rise of design thinking in the early 2000s, anticipatory design holds the potential to become a core strategic tool—one that drives innovation not just through better functionality but by aligning systems more deeply with human goals. The opportunity is vast, but delivering on it requires more than visionary thinking. It demands methods that allow us to move from speculation to implementation.

10.2.1 Why Mixed Methods Matter for Anticipatory Design?

To translate anticipation into design practice, we need tools that can handle uncertainty, complexity, and nuance. That's where mixed methods come in. Designing anticipatory systems isn't about choosing between intuition and data—it's about integrating both. A mixed-methods approach combines the **richness of qualitative insights** with the **rigor of quantitative analysis**, helping designers build imaginative and grounded systems.

Qualitative methods are invaluable at the exploratory stage. They help uncover user values, contextual nuances, and emerging behaviors—especially in ambiguous or evolving domains. Techniques like scenario planning, environmental scanning, and trend analysis reveal possible futures and help designers frame the right questions before jumping to answers.

Quantitative methods, on the other hand, bring precision and scalability. They help structure and test hypotheses, simulate possible outcomes, and generate predictive models based on real-world data. Tools like forecasting models, PA, or decision trees allow designers to evaluate how systems might perform across a range of variables and conditions.

Used together, these methods offer more than the sum of their parts. Qualitative insights can inform model design, ensuring that predictions reflect real human behavior, not just patterns in the data. Quantitative validation, in turn, adds confidence and credibility to insights that begin in conversation, observation, or speculation.

This integrative approach isn't just helpful—it's essential. Anticipatory systems must balance **what could happen, what should happen,** and **what people want to happen**. That requires sensitivity to both human narratives and algorithmic logic. Mixed methods provide the scaffolding for this kind of design—one that's imaginative yet evidence-based, flexible yet responsible.

As you've seen, anticipatory design isn't just about building more intelligent systems—it's about designing systems that *care*. Systems that learn, evolve, and respond in supportive, not manipulative ways. And while this vision is inspiring, it's also incredibly complex. Designers need ways to explore uncertainty while staying grounded in evidence. That's why mixed methods matter. They offer a practical bridge between empathy and analytics—between stories and statistics. In the next section, we'll explore a selection of qualitative and quantitative methods that can help designers bring anticipatory systems to life, responsibly and effectively.

10.3 MIXED METHODS: DESIGNING FOR THE FUTURE WITH BOTH HEART AND MATH

I will delve deeper into these topics later in the book, but it's beneficial to highlight a few methods now [51]:

10.3.1 Qualitative Methods

- **Trend analysis** provides insights into long-term patterns and cycles. However, may overlook abrupt changes or emerging trends.
- **Scenario planning** promotes strategic thinking and readiness for diverse futures. Yet creating multiple scenarios requires significant collaboration, time, and resources.
- **Delphi method** utilizes the knowledge of a varied group of experts but may be vulnerable to bias or groupthink.
- **Environmental scanning** helps organizations monitor external influences, but it can be difficult to filter and prioritize the overwhelming volume of information.
- **SWOT analysis** is an effective and straightforward tool for identifying internal and external factors impacting an organization. However, it might be subjective, resulting in biased results.
- **Qualitative risk assessment** offers a comprehensive perspective on potential risks, including those that are hard to quantify. Still, it relies significantly on subjective judgments, which can differ among stakeholders.
- **Brainstorming** encourages divergent thinking but requires facilitation to remain productive; otherwise, it may result in unfocused discussions or impractical ideas.

10.3.2 Quantitative Methods

- **Time series analysis** uses historical data to identify trends and patterns. However, it assumes that past trends will continue, which may not always be true.
- **Regression analysis** reveals the relationships among variables for predictive insights. However, it requires assumptions about the nature and consistency of these relationships, which may not always be accurate.

- **Monte Carlo simulation** models uncertainty through probabilistic results. It requires a solid understanding of probability distributions and input variables.
- **Forecasting models**: Provide organized and methodical ways to anticipate future trends. However, accuracy heavily depends on the quality of historical data and the suitability of the selected model.
- **Decision trees** illustrate decision-making processes and possible outcomes. They might simplify intricate decision contexts and the interactions between variables.
- **Bayesian analysis** integrates prior knowledge and updates beliefs based on new evidence. It relies on reliable prior data, which may be subjective and affect results.
- **Predictive Analytics** uses advanced algorithms to extract insights and forecast behaviors. Requires large, clean, and structured datasets, as well as expertise in data preprocessing and algorithm selection.

10.3.2.1 Bringing It Back to Design

Designers don't need to master all these methods, but understanding their roles can make collaboration more effective and strategy more grounded. A designer leading an anticipatory service redesign can

- Use **scenario planning** to define plausible user futures.
- Collaborate with data scientists using **Predictive Analytics** to map user behaviors.
- Use **environmental scanning** and **trend analysis** to feed design explorations.

By becoming fluent in these mixed methods—or at least comfortable navigating them—designers can help shape anticipatory systems that are not just reactive to change but thoughtfully prepared for it. Remember, in anticipatory design, the best future-forward solutions are grounded in today's evidence and tomorrow's imagination.

As you've seen, anticipatory design is not a single method or mindset, but a landscape of perspectives, tools, and possibilities. Whether through speculative foresight or rigorous system modeling, the goal remains the same: to design systems that care, guide, and adapt. In the next chapter, we'll begin to explore how these principles take shape when applied to real-world contexts, where design meets uncertainty and futures begin to unfold.

TL;DR

Design schools of thought—Anticipatory design is still an emerging field, shaped by different schools of thought, each offering unique priorities:

- **Celi and Colombi**: Use speculative design and trend-spotting to imagine multiple futures and cultural shifts.

- **Cascini**: Uses engineering principles to foresee system evolution and enhance future-ready systems—technically rigorous, yet less focused on people.
- **Morrison et al.**: Apply a critical lens, focusing on ethics, sustainability, and systemic relationships across scales.

These perspectives are valuable for expanding our thinking, but often remain abstract or complex to apply directly to product and UX design.

Why this matters for designers?—If we build systems that understand why people behave the way they do—and how that might change—we can design tech that:

- Focus on **user intent and decision-making** as the core drivers of responsible anticipatory systems.
- Bridge qualitative and quantitative methods—using trend analysis, scenario planning, and predictive modeling to build trustworthy, context-aware experiences.
- Avoid over-automation by prioritizing transparency, adaptability, and ethical foresight.

Anticipatory Design

11

Designing for What Comes Next

Suppose we want to build technology that feels truly intelligent and helpful. In that case, we must shift our focus from supporting immediate tasks to designing for long-term outcomes—aligning systems with what users ultimately seek to achieve, not just what they're doing at the moment. That shift defines anticipatory design. This chapter marks a turning point. So far, we've explored anticipation through various disciplinary lenses and laid out how anticipation influences human behavior and systems. Now, we begin the practical work: how do these ideas translate into real products, services, and user experiences? How can we design systems that act not just in response to human input but in alignment with human intent?

As discussed in earlier chapters, anticipatory design leverages behavioral and contextual data to predict needs and reduce friction. But turning this promise into practice requires more than data—it demands thoughtful design methods. This chapter explores how to operationalize those capabilities, aligning short-term actions with long-term intent to build systems that support, guide, and adapt alongside their users.

This marks a fundamental shift from reactive systems that wait for user input to intelligent services that act in advance to support users more intuitively. Traditional digital experiences often rely on a reactive model: the user initiates, and the system responds. In contrast, anticipatory systems can act on behalf of the user, using resources like big data (BD), ML, and AI. This simplifies decision-making, reduces bias, and eases cognitive overload. As Shapiro puts it, these systems aim to "respond to user needs one step ahead of their decisions" [27].

DOI: 10.1201/9781003642800-14

11.1 ANTICIPATION IN ACTION: FROM CONCEPTS TO CAPABILITIES

Anticipatory design depends on two key capabilities:

- **Prediction (past→present)**: Using historical and contextual data to estimate what a user might do or need next.
- **Foresight (present→future)**: Looking further ahead to anticipate how a user's needs might evolve—and designing with that future in mind.

Consider Amazon's recommendation engine: it utilizes your past purchases to suggest products you might be interested in now. This process exemplifies prediction. On the other hand, envision a smart home system that observes and learns your evening routine over time. It not only recognizes when you typically turn on the lights or adjust the thermostat but also starts to automate these actions, fine-tuning energy consumption to align perfectly with your habits. This capability reflects foresight—the ability to anticipate user needs based on learned behaviors. Together, prediction and foresight enable us to shift from merely responding to user requests to actively anticipating their needs even before they become conscious of them.

11.1.1 The Evolution of Digital Services: From Reactive to Anticipatory

Over the last two decades, digital services have steadily moved from reactive interfaces to predictive personalization. Early systems responded only to direct input—users clicked, and the system served results. Personalization added a layer of intelligence, tailoring content based on past behavior. But even that often relied on static models and narrow assumptions.

We're now seeing a broader shift toward services that not only adapt to users' actions but also anticipate what they're likely to want or need next. This evolution introduces a new design responsibility: moving beyond historical data to include real-time signals, context, and evolving intent. Designers must now consider not just how to react quickly but how to act meaningfully, not just how to reduce effort but how to design for relevance in moments that matter. This is where anticipatory design departs from traditional UX—it's not just about designing for interaction but about designing for alignment between system behavior and human purpose.

We've seen how **prediction** helps systems react to immediate signals while **foresight** allows them to design for longer-term possibilities. But how do we operationalize these capabilities in real-world systems? That's where **forecasting** enters the picture. **Forecasting equips systems to transition from reactive to proactive behavior.** It's the process that bridges data-driven prediction and human-centered foresight. It allows

Reactive

Guide user make decisons

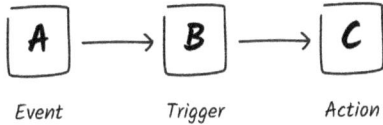

| A | → | B | → | C |
| Event | | Trigger | | Action |

Proactive

Make decisions on behalf of user

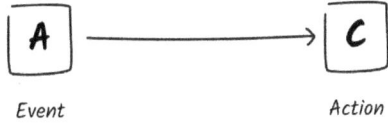

| A | ⟶ | C |
| Event | | Action |

FIGURE 11.1   My understanding of how experiences transition from reactive to proactive.

systems to estimate likely outcomes based on historical and contextual data while also accommodating uncertainty and change. In this way, forecasting becomes a foundational capability of anticipatory design. It enables us to move from designing for what users are doing now to what they are working toward next. This shift marks a new era in design, moving from experiences that respond only when prompted to ones that proactively serve users in real time.

Current digital products are often still reactive: a user does something, and the system responds (A → B → C). In this setup, the agent, meaning the user, regulates and controls their actions to reach a result. However, this constant demand for user input can be overwhelming, contributing to decision fatigue or even analysis paralysis [52]. Anticipatory systems aim to shift this by skipping unnecessary steps—jumping from A → C—where the system begins to regulate and support the user's actions based on its perception of what's likely to happen next. This leap is only possible when systems have the capacity to model intent, not just past behavior. In other words, the system becomes proactive, helping users move toward their goals without waiting for explicit instructions [52].

Designing systems like this isn't easy. A Forrester survey of over 350 Business-to-customer (B2C) companies revealed that many still struggle to anticipate customer needs effectively [53]. Why? Because doing it right means deeply understanding user behavior and having the real-time capabilities to act on it. Anticipatory design depends on learning systems that continually adapt by collecting and analyzing user data, predicting intent, and refining how they respond.

In anticipatory systems, complexity is part of the process. These systems manage large volumes of personal, real-time information. They don't just learn—they evolve, sometimes developing new features based on what users need before they've even asked. This shift also means traditional success metrics might not always apply. For instance, a drop in user interactions might actually reflect a smoother, more intuitive experience.

Anticipatory design supports the user throughout their journey:

- **Before** the interaction begins, by surfacing what they might need.
- **During** the interaction, by offering context-aware suggestions.
- **After** summarizing insights and simplifying decisions.

Predicitive

Human-centered Design

Anticipated

Intent-driven Design

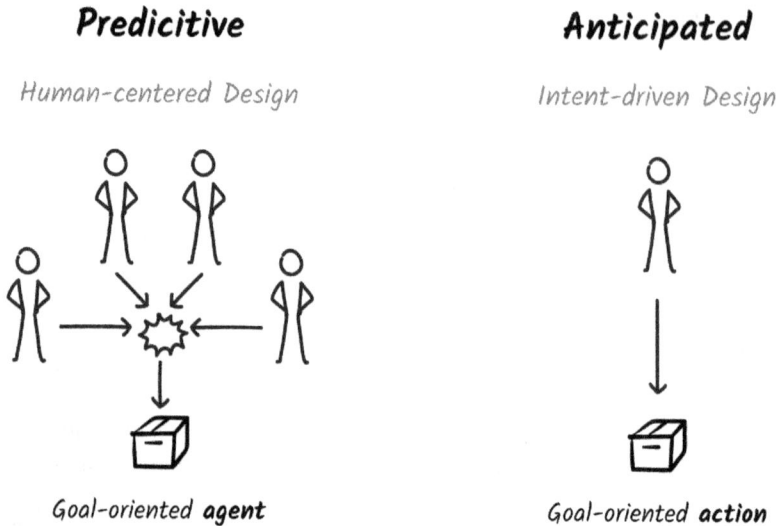

*Goal-oriented **agent***   *Goal-oriented **action***

FIGURE 11.2   Moving from predictive to anticipated systems.

The result? Less mental effort and smoother, more efficient interactions. To understand how this works, think of it in terms of behavioral patterns:

- Reactive systems follow a simple formula: **Stimulus → Action (S-A)**.
- Anticipatory systems take it further: **(Stimulus + Expectation) → Action (E-A)**.

These systems can even work backward: **Stimulus → Expectation (S-E)** or **Stimulus + Action → Expectation (S-A-E)**. This shift—where a system predicts what's likely to happen and acts accordingly—is what separates prediction from true anticipation [31].

I anticipate this challenge for the future of anticipatory design research: the passage from predictive systems toward anticipatory systems. The constitutive definition of anticipation indicates a state where something is expected, and action is taken toward that expectation.

Here's the key difference:

- **Predictive systems** rely on hindsight. They look backward to guess what comes next.
- **Anticipatory systems** use foresight. They look forward and adjust the present accordingly.

This shift has profound implications. In predictive systems, the user remains in control, driving interactions and making choices. In anticipatory systems, the system starts to guide, recommend, and even act on the user's behalf. It doesn't just wait to be told what to do—it uses what it knows about us and what might happen next to offer support in real time. Here's how that shift looks when broken down:

This shift—from **reactive support** to **proactive partnership**—marks one of the most critical changes in digital design today. It's not just about predicting what a user might do. It's about aligning system behavior with their future goals—sometimes before those goals are fully formed in the user's mind. As we progress, we'll design more systems that support users in achieving their goals, not just executing commands. This represents a shift from **goal-oriented agents**, where the human regulates and controls their actions to achieve a result, to **goal-oriented actions**, where the system supports and governs the user's actions based on its perception of expected outcomes [31].

TABLE 11.1   Comparison Between Predictive and Anticipatory Design Approaches.

	PREDICTIVE	*ANTICIPATORY*
Focus	The system waits for user input (goals, decisions, data).	The system proactively takes action on the user's behalf.
Process	Users act as agents who interact with the system, and the system reacts based on predictions.	It infers user intent and delivers value without explicit input.
Design logic	The system analyzes *past behavior* or current input to recommend or suggest next actions.	Rooted in *intent*, context, and foresight, aiming to reduce user friction.
Goal-oriented	**Agent**: The human initiates actions, and the system supports or predicts based on them.	**Action**: The system becomes the agent, acting *for* the human based on inferred intent.
Use case	A recommender system that suggests movies based on what you watched.	A calendar assistant that reschedules a meeting before you ask, based on travel delays.

TABLE 11.2   Distinct Features of Predictive and Anticipatory Systems.

	PREDICTIVE SYSTEMS	*ANTICIPATORY SYSTEMS*
Strategy	Looks backward (hindsight) to make predictions	Looks forward (foresight) to anticipate needs
Focus	Supports the user after they act	Helps the user before they act
Control	The user controls the system through direct input	The system supports the user by anticipating intent
Type of Goal	**Implicit**—the system reacts to patterns without fully understanding your broader intent	**Explicit**—the system has a clearer understanding of your future goals and acts accordingly
Data Type	**Past → Present** **What powers it is** historical data and behavioral patterns	**Present ← Future** **What powers it is** real-time context and predicted outcomes
How it Helps	Suggests or recommends based on past choices	Guides, adjusts, and acts based on where the user is likely headed next

11.1.2 From Generation to Agency: Where Anticipation Begins

While we often assume that AI solutions should be proactive, the truth is that many tools we use daily, including large language models (LLMs) for language and image generation, are actually reactive. Therefore, it is crucial to pause and understand the key characteristics that set anticipatory systems apart. To clarify, we should distinguish between two developing AI paradigms: **Generative AI** and **Agentic AI.**

Generative AI—like chatbots or image generators—is inherently **reactive**. It waits for a prompt, and then generates content based on patterns learned from training data. Its job ends when the output is delivered. It doesn't initiate action or evolve its strategy without human input.

Agentic AI, by contrast, is **proactive**. It may still begin with a prompt, but it continues beyond generation—it plans, acts, evaluates results, and adapts. Think of a personal shopping assistant that monitors prices, compares vendors, and completes purchases with minimal intervention. It perceives, decides, acts, and learns, often using generative AI as a reasoning engine.

This distinction is important. When designed effectively, anticipatory systems **lean toward agentic behavior**. They don't simply suggest—they **take initiative**. They don't just output—they **engage in goal-driven action**. Most importantly, they **learn from outcomes**, improving their ability to serve user intent over time.

As AI evolves, anticipatory systems will increasingly blend generative and agentic capabilities. The goal isn't to generate more—it's to **guide better**, by combining foresight, planning, and real-time adaptation. Therefore, to move from reactive support to proactive partnership, systems must be able to estimate what users need, sometimes even before users know it themselves. But anticipation doesn't happen by accident. Structured methods are required to help designers think ahead and act with intention. That's where forecasting and backcasting come in.

11.2 DESIGNING WITH THE FUTURE IN MIND: FORECASTING VERSUS BACKCASTING

Forecasting is the engine of prediction. It analyzes historical trends to anticipate future behavior. Netflix, for example, uses forecasting to suggest your next favorite show based on what you've already watched. But forecasting has its limits. It relies heavily on historical data, presenting a significant challenge for new users with little to no interaction history, a phenomenon that was already explored in Chapter 1, the *cold start* problem.

This is where **backcasting** becomes critical as a complementary methodology. Instead of predicting what might happen based on the past, backcasting starts with

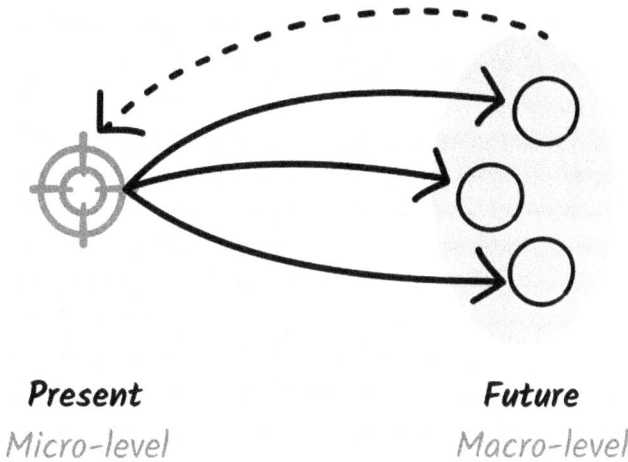

Present

Micro-level

Future

Macro-level

FIGURE 11.3 Visualizing the relationship between forecasting and backcasting in anticipation design.

a clear goal and works backward to determine how to reach it. By using backcasting, designers can establish system behavior aligned with long-term user goals, even without rich historical data. For example, imagine a financial app helping users save for retirement. Forecasting looks at spending patterns to suggest small changes. Backcasting starts with the user's long-term goal, for instance, retiring at the age of 54, and builds a path backward to meet it.

These two methodologies must work together to achieve anticipatory experiences successfully: the large circle represents future goals (backcasting), and the arrows show how past trends guide the way toward these goals (forecasting).

Forecasting and backcasting represent the **dual forces of circularity in anticipatory design**: the future informs the present, and the present shapes the future. It implies circularity [54] (we will return to this topic in just one moment). Consequently, how can we design a future state of an experience that affects the present time?

11.2.1 Micro- and Macro-Perspectives

Forecasting and backcasting don't just differ in direction—they operate at different levels of analysis. Forecasting typically operates at the microlevel. It zooms in on individuals, modeling their behavior through short-term signals and contextual patterns. That's why behavior science frameworks like B=MAP, TTM, and Nudge theory are so valuable: they help designers craft timely, personalized interventions based on current motivation, ability, and behavioral readiness to inform the present.

In contrast, backcasting works at the macro level. It begins with a long-term outcome—whether individual (e.g., better health), societal (e.g., net-zero emissions), or organizational (e.g., inclusive design)—and works backward to identify what must change today to achieve it. Behavior frameworks like **COM-B** (Capability, Opportunity, Motivation = Behavior) or foresight tools like **Futures Cones** are essential here. They allow designers to model broader systems of change—policy, culture, environment—and map how individual behaviors intersect with these systemic forces. These tools are handy for addressing complex or "wicked" problems where user behavior is influenced by infrastructure, legislation, or social norms.

By layering **micro level forecasting** with **macro level backcasting**, designers can align day-to-day system interactions with long-term transformation. For instance:

- A **micro level intervention** might prompt a user to walk 10 minutes after lunch (nudging healthier routines).
- A **macro level strategy** might involve designing a city-wide step-tracking initiative supporting urban mobility goals and accelerating infrastructure investments.

Thus, anticipatory systems go beyond mere convenience and toward responsibility, fostering both individual agency and collective foresight. Design serves as a means to transform future-oriented objectives into significant actions in the present, striking a balance between detailed support for individuals and broader guidance for systemic change.

11.2.2 Aligning Present Actions With Future Intent

As user expectations for digital experiences continue to evolve, there's a noticeable shift toward systems that proactively meet needs rather than respond. This shift increases the importance of combining both forecasting and backcasting because forecasting will

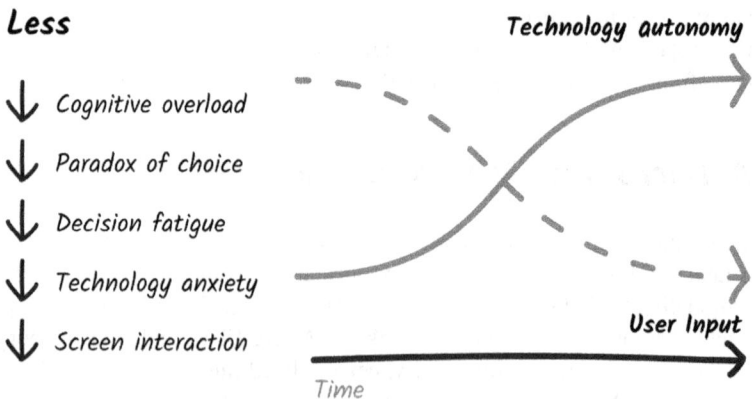

FIGURE 11.4 A correlation exists between increased automation capacity and reduced user cognitive overload.

focus on anticipating short-term behavior by identifying patterns and trends, and back-casting will ensure that those patterns align with bigger-picture user outcomes. Together, they guide near-term actions while anchoring them to long-term intent. Designers and data scientists must collaborate using these two complementary methods to create these forward-looking solutions.

Research suggests a positive correlation between increased automation capacity, when driven by accurate forecasting, and reduced cognitive overload [55, 56].

By anticipating users' needs and behaviors, forecasting empowers anticipatory systems to [57]:

- **Personalize more effectively**: Deliver hyper-relevant recommendations and information.
- **Provide proactive assistance**: Offer assistance before user needs are explicitly stated, streamlining user journeys.
- **Reduce cognitive overload**: Present relevant options at the right time, minimizing mental effort.
- **Improve decision-making**: Predict potential outcomes based on user history and similar patterns.
- **Minimize errors**: Foresee potential issues, reducing frustrations and mistakes.

11.3 APPLICATIONS OF CIRCULARITY IN DESIGN

To design for what's next, we must understand what's happening now and why it's happening. That's where **causality** comes in. Circularity refers to the feedback loop between what's happening *now* and what might happen *next*. There are two key types:

- **Forward circularity**: The system anticipates a future event and acts in the present to influence that outcome, taking proactive steps based on expected developments.
- **Backward circularity**: The system monitors current behavior to refine its understanding of future needs, adjusting its predictions based on how the user engages with it in the present.

Anticipatory systems don't just react. They analyze cause-and-effect relationships to predict outcomes. They must ask: what happened before? What tends to follow? How does this behavior relate to that goal?

Unlike reactive systems, which just respond to what users do, anticipatory systems try to figure out *why* users do what they do. They look at the cause-and-effect relationships behind our behavior to better predict what we might need in the future.

Consider a smart home system, for instance. It doesn't simply turn on the lights when instructed. It monitors that every day at 7:30 pm, you switch on the reading lamp and lower the temperature as the room grows darker. Over time, it learns the *cause* (low light, time of day) and the *effect* (you desire a cozy reading environment). Instead of waiting for prompts, it begins to provide that experience proactively.

11.3.1 *What-Ifs*: The Power of Counterfactual Reasoning

But understanding what causes what isn't always enough—systems must also explore alternative futures. This is where counterfactual reasoning becomes a powerful tool. This goes beyond just pattern-matching. Great anticipatory systems ask, *What would have happened if. . .?* This is called **counterfactual reasoning** [58]. It's how systems test alternative outcomes to learn what truly matters.

Think about a financial app helping a user plan for retirement. It might simulate what happens if they save more, spend less, or retire earlier. Each "what-if" scenario helps users better understand the long-term impact of their decisions—*and* helps the system refine its guidance based on what works best.

11.3.1.1 Causality Isn't Always a Straight Line

Most of us are used to thinking of causality as linear: *A causes B*. But in complex systems—like cities, ecosystems, or people's daily lives—cause and effect are often more circular. **Your actions shape outcomes, but those outcomes also shape your future actions.** This feedback loop, called **circularity**, is essential for designing adaptive, future-facing systems [54].

Circularity in anticipatory systems refers to a dynamic, two-way relationship between present actions and future outcomes. In what's called *forward circularity*, future scenarios shape present decisions, like when a navigation app adjusts your current route based on predicted traffic ahead. In **backward circularity**, present actions reshape future expectations, just as a climate policy needs to adapt as results come in. This continuous feedback loop allows anticipatory systems to remain adaptive and responsive, with the future and present constantly informing one another.

11.4 WHY THIS MATTERS FOR DESIGNERS?

Understanding causality and circularity helps us build smarter, more adaptive, and more human-centered systems. To design for what's next, systems must understand both what users do and why they do it. That means:

- Better personalization that respects the intent
- Smarter recommendations that adapt in real time
- Systems that grow more helpful over time

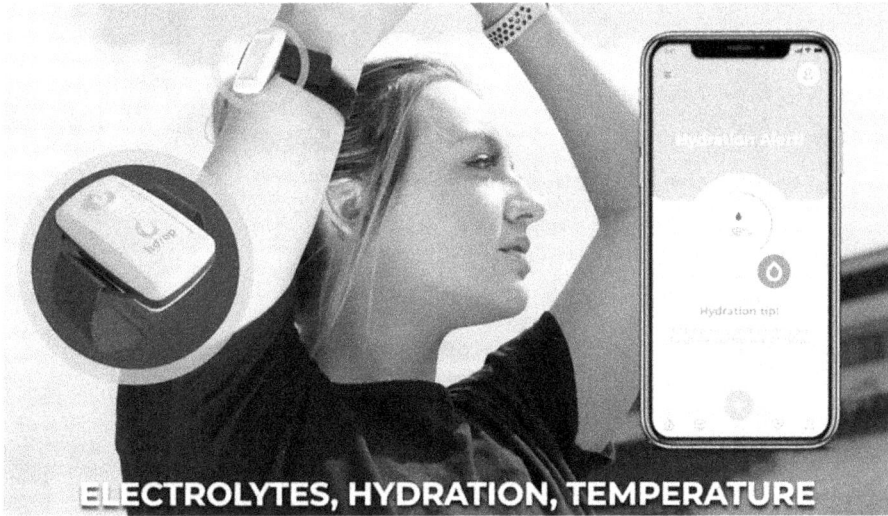

FIGURE 11.5 Representation of hDrop, a noninvasive device that claims to assess hydration and electrolyte levels accurately [59].

In short, **causality turns data into insight**, and **anticipation turns insight into action**. Together, they help us design technology that truly *feels* intelligent, not because it's magic but because it understands how and why people move through the world. Let's look at how these ideas come to life in practice.

Forward circularity means that the system uses potential *future* scenarios to guide present actions. A great example of this is **hDrop**, a wearable device that predicts hydration levels. Instead of merely reacting when the user is dehydrated, hDrop anticipates needs, factoring in ambient temperature, past activity, and even upcoming exercise, to recommend hydration before the issue arises. The system simulates possible futures and acts accordingly, illustrating how prediction, action, and user goals can align.

On the other hand, **backward circularity** involves using present actions to *update* future expectations. Consider **Absorb**, an intelligent learning management system (LMS). As learners engage with course content, the system adjusts its curriculum in real time, identifying patterns in areas where users struggle and adapting the path forward. It doesn't merely predict which content might be useful; it learns from user interactions and evolves to remain relevant. This adaptive loop ensures that each new interaction informs the next one.

Ultimately, circular thinking manifests in **scenario planning and backcasting** techniques that assist organizations in aligning their long-term objectives with immediate actions. For example, a ride-sharing company may foresee a future dominated by an electric fleet due to emerging emissions regulations and evolving consumer preferences. From that envisioned future, it then retraces its steps: what infrastructure investments should be initiated now? Which collaborations must be established? What adaptations

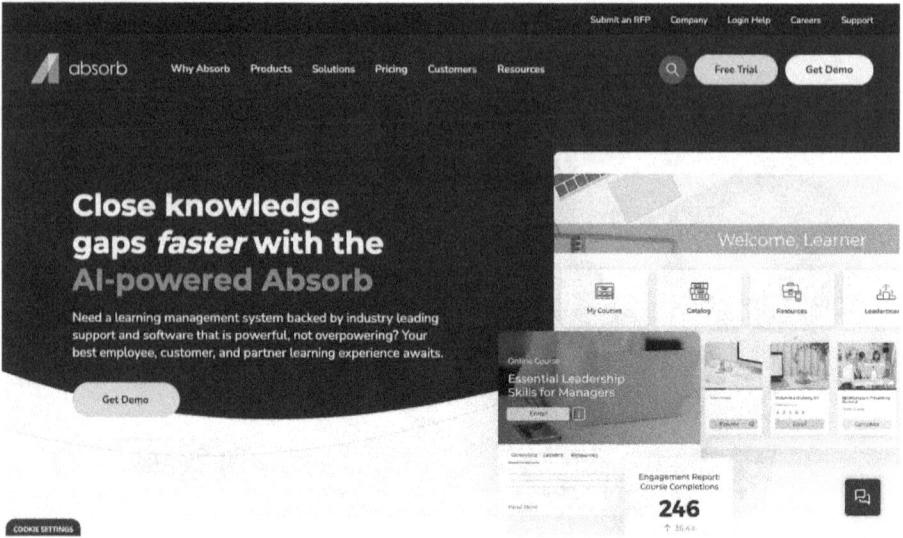

FIGURE 11.6   Screenshot of the www.absorblms.com website in 2025.

are necessary for the user experience? In this context, circularity ensures that today's choices consider the realities of tomorrow.

These examples have a common characteristic: they view the *future* as a vital component of the *present*. By incorporating circularity and causality into our designs, we can develop systems that are not only smart but also robust, adaptable, and in better harmony with human behavior.

11.5 ANTICIPATION CAPABILITIES: IMPLICIT VERSUS EXPLICIT

As previously introduced, Roberto Poli frames anticipation in two distinct modes: **implicit** and **explicit** [30].

Implicit anticipation refers to fast, intuitive responses—those automatic reactions our minds and bodies have learned through experience. Imagine catching a falling object or stepping back from a car speeding toward you. These anticipatory actions happen almost without thought, shaped by past exposure and environmental cues.

Explicit anticipation, on the other hand, is slower and more deliberate. It involves planning, forecasting, and modeling future possibilities. This kind of anticipation underlies long-term decision-making, such as preparing a career path, designing a product roadmap, or simulating future scenarios based on predictive data.

Both types share a common foundation: they generate imagined futures to shape present action. One operates through instinct, the other through strategy, but both are crucial to how humans (and systems) navigate uncertainty. In the context of anticipatory design, these modes translate into different levels of decision support. Anticipatory systems can help users by [60]:

- **Simplifying selections**—Reducing the effort needed to make a choice through intelligent defaults or pre-filled options.
- **Choice editing**—Narrowing down options based on context, preferences, or behavioral patterns.
- **Eliminating decisions**—Automatically handling low-stakes choices when user intent is clear—removing the need for the user to make a decision at all.

Together, these *strategic capabilities* define how anticipatory systems intervene in decision-making, shaping the level of effort, control, and relevance users' experience.

11.5.1 Simplifying Selections

ML enables algorithms to create customized pre-selection defaults and retain user preferences for personalized service. Spotify's personalized features, especially "Your Daily Drive," utilize sophisticated ML algorithms to provide users with precise and customized content. When Spotify recommends a playlist, it doesn't rely on chance but instead makes an informed prediction based on a user's listening history and tastes, anticipating the types of songs they enjoy.

Spotify applies advanced ML to vast datasets—tracking what you play, skip, and save—to deliver playlists tailored to your preferences. A 2017 study by The Echo Nest (which Spotify acquired in 2014) confirmed the accuracy of these predictions and their

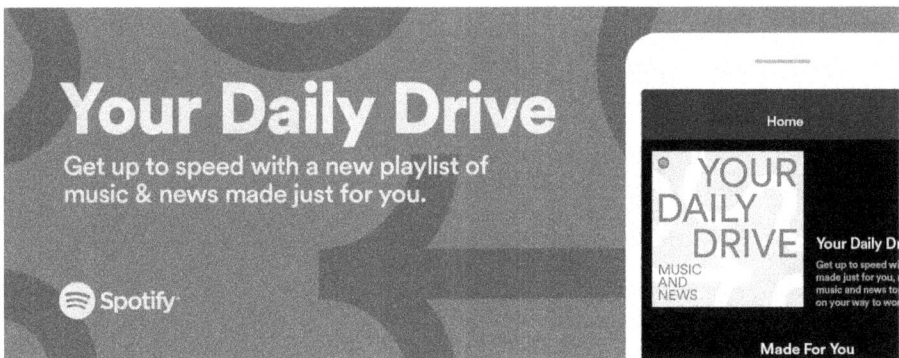

FIGURE 11.7 An example of anticipatory design is Spotify's playlist suggestion "Your Daily Drive" [61].

impact on user satisfaction. Because these recommendations often feel spot-on, users avoid the effort of building playlists themselves. The outcome isn't just convenience— it's reduced decision fatigue, as the system does the heavy lifting of choice, quietly aligning with your taste.

11.5.2 Choice Editing

Curation plays a critical role in anticipatory design, especially regarding editing choices down to what truly matters. By removing irrelevant or overwhelming options, systems can help users make decisions more quickly, confidently, and with less mental effort.

One compelling example is **Easysize**, an AI-powered clothing company that uses contextual data to deliver more intelligent sizing recommendations. Rather than asking users to navigate complex sizing charts, Easysize anticipates fit by analyzing collective behavioral patterns, purchase history, product data, and user feedback. The result? Fewer wrong-size orders—and fewer returns.

Sizing issues are not a small problem. The high percentage of returns due to incorrect sizing represents a significant cost to the fashion industry and the environment. According to Easysize, 30–40% of all online fashion is returned due to a misfit. Worldwide, this amounts to an astonishing $32 billion (about $98 per person in the United States) on handling size-related returns annually. This represents a substantial logistical effort and cost for brands and even for the sustainability of our planet, given that the items must move around from one place to another. By narrowing options to fit best, Easysize reduces decision fatigue and supports more sustainable consumption.

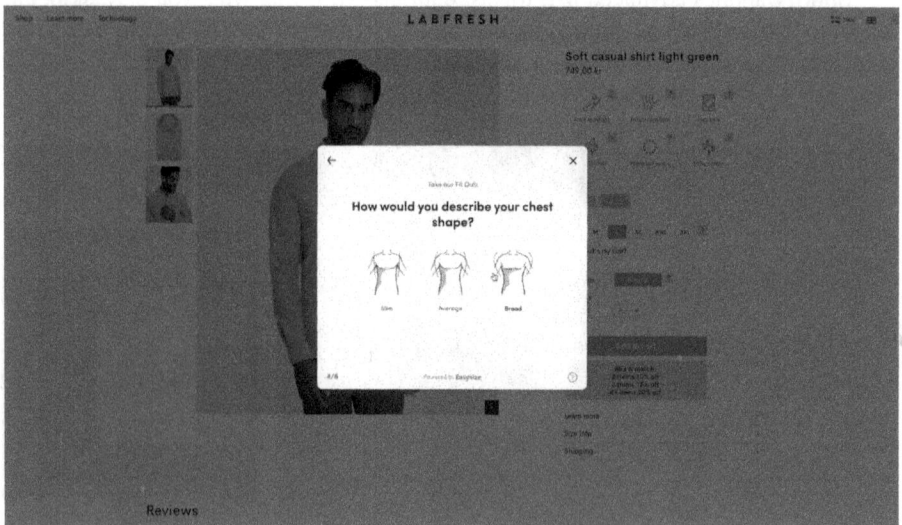

FIGURE 11.8   An example of anticipatory design is Easysize, an AI-powered size recommendation service [62].

Their strength lies in a lot of contextual data. And that's the broader lesson for anticipatory design: effective choice editing isn't about removing options blindly— it's about using contextual intelligence to remove the noise and highlight what's most relevant.

11.5.3 Eliminating Decisions

One of anticipatory design's most ambitious goals is eliminating the need for certain decisions altogether—ideally without users noticing. When done well, this can lead to delightful, frictionless experiences where systems handle low-stakes, repetitive choices on behalf of the user. But this is also the most delicate form of anticipation. Users can feel disempowered, surveilled, or trapped in rigid systems if too much control is removed. The challenge is to eliminate effort without eliminating user agency.

Achieving this level of anticipation requires systems that can interpret context, model behavior, and adapt to change—all while respecting user intent. Designers working in this space must cultivate anticipatory thinking skills: reading system states, forecasting likely outcomes, and planning for alternate futures. This approach calls for a human-centered AI mindset, where algorithms don't just act autonomously—they act accountably, ethically, and transparently.

One domain where decision elimination has found early success is personal finance. Apps like **Qapital** use automation not to replace the user but to support them, removing

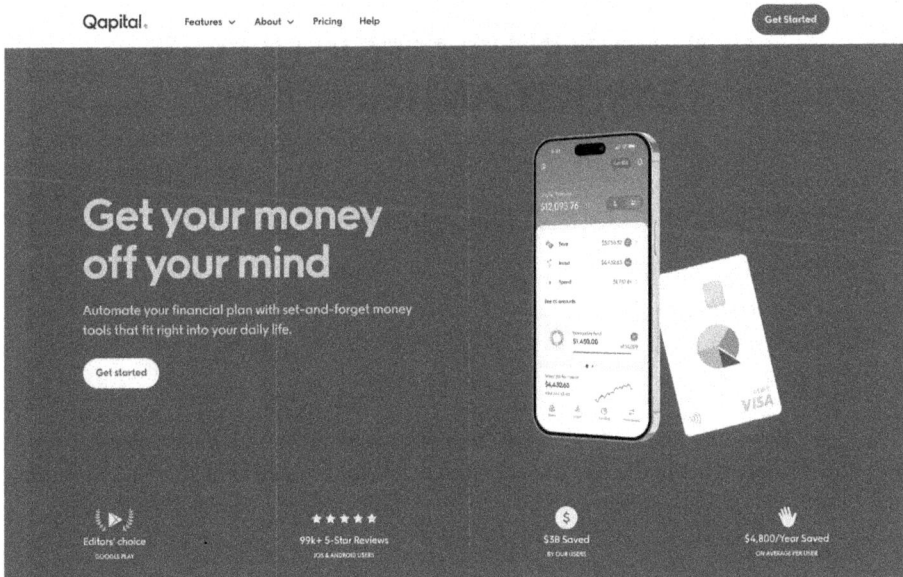

FIGURE 11.9   Screenshot of Qapital app website as it appears in 2025.

Source: qapital.com

the cognitive burden of everyday money decisions while keeping long-term goals visible and intact. For example, users can set savings rules like "round up every purchase" or "save $5 every time I skip takeout." Once the rules are in place, the system quietly does the work in the background. Users don't have to make daily savings decisions—they just watch their goals get closer with minimal effort.

Qapital's strength lies in how it combines automation with agency. Users retain control over what gets automated, but once the system is set in motion, it performs reliably and transparently. This creates a relationship of **proactive support**, where the system fulfills the user's goals, even when the user isn't actively engaged. It's anticipatory design at its most invisible yet impactful. Of course, eliminating decisions doesn't mean removing control. Designers must be cautious. Eliminating decisions without clear communication or opt-out options can lead to mistrust. In high-stakes domains like health or finance, anticipatory systems must provide **feedback loops**, allow **user override**, and respect **individual autonomy**. This calls for a design mindset rooted in human-centered AI—one that not only anticipates needs but also anticipates the limits of automation.

Designers can apply similar principles across domains. Whether you're creating financial tools, wellness platforms, or smart home systems, the key is understanding what decisions *users want help with* and which ones they *still want to own*.

Eliminating decisions isn't just about convenience—it's about care. When anticipatory systems quietly remove noise, they leave behind something more meaningful: space for users to focus on what truly matters.

11.6 THREE STRATEGIC USES OF EXPLICIT ANTICIPATION

To accomplish that, designers can use explicit and implicit anticipation in different ways to explore new systems and needs. Within explicit anticipation, there are three main use cases [63]:

1. Optimization
2. Contingency
3. Novelty

The point of distinguishing these three categories is to assist with the challenge of linking specific tasks to specific methods or approaches for both thinking about and shaping the future.

11.6.1 Optimization

Optimization represents the clearest example of explicit anticipation. It's the intersection of design and data, where past successes inform future developments. Through optimization, we aren't creating completely novel concepts. Instead, we focus on enhancing, adjusting, and refining existing ideas. This approach relies on the belief that historical trends can effectively inform future actions. This could involve leveraging behavioral data to tailor content, enhance user interactions, or minimize complications and errors.

Optimization frequently manifests as algorithmic modeling—systems leveraging historical data to produce future outcomes. This approach is based on the assumption of continuity: that trends will remain constant, preferences will persist, and enhancements will blend over time.

Optimization not only is powerful but also carries a risk: it can unintentionally reinforce the status quo. If we focus solely on past outcomes to inform future designs, we risk excluding novelty, ignoring edge cases, and perpetuating bias. When used wisely, optimization brings stability and performance; when applied blindly, it can limit our vision of what's possible.

11.6.2 Contingency

Contingency planning occurs when we step outside of what's likely and prepare for what's possible. Unlike optimization, which builds from established patterns, contingency planning prepares for disruption. It asks, *What if things don't go as expected?*

This form of explicit anticipation is all about readiness. It doesn't try to predict a single outcome—it anticipates a range of surprises, from subtle deviations to major shifts. Contingent thinking is especially valuable in volatile contexts, where external forces—economic shifts, climate events, social movements—can change everything.

In design, contingency means building systems that can adjust to pressure: adapting to unpredictable user behavior, adjusting to environmental changes, or offering fallbacks for unexpected edge cases. Designers may simulate potential failures, evaluate alternative routes, or stress-test interactions and experiences for resilience. They may also work closely with developers to define and evaluate reward functions that help systems learn from surprises without drifting into unintended behavior (a topic we first explored in Chapter 4). When carefully crafted, these functions allow systems to anticipate misalignment and gently course-correct rather than collapse.

This mindset improves our ability to track uncertainty in real time, learning from the unexpected, as it reveals. Designers can use contingent futures to plan how a system responds to surprises, especially when those surprises carry risk. For example, they can design and evaluate the reward function.

11.6.3 Novelty

If optimization looks to the past and contingency prepares for surprise, novelty embraces the unknown. It's the most imaginative and demanding form of explicit anticipation—one that invites designers to explore openness rather than certainty.

Novelty futures ask: *What could exist that doesn't exist yet?* These are futures without precedent, where we invent entirely new ways of seeing, doing, or being. In design, novelty is sparked through speculative tools like design fiction, counterfactual scenarios, or radical prototyping. These methods help us break free from linear thinking and open space for genuine innovation.

But designing for novelty is not about guessing wildly. It's about developing the cognitive and ethical flexibility to imagine systems that can adapt to futures we can't yet fully describe. It's about equipping ourselves to respond not only to user needs but also to societal shifts, environmental crises, or cultural transformations. In short, novelty challenges us to anticipate not just what users might want, but who users might become.

There is a general agreement that these three ways of using the future under explicit anticipation do not live in isolation; all are usually employed in different proportions [64, 65]. Nevertheless, differentiating between these categories helps designers and non-designers shape the future through specific methods and approaches, creating sustainable and trustworthy anticipatory experiences. The goal is to analyze how people are "using the future" to make it easier to match tools to tasks. This will help establish an anticipatory system in four ways:

1. Accelerating the pace of change in a high level of uncertainty.
2. Engaging and facing wicked problems and existential threats.
3. Assuming shared responsibility for positive outcomes.
4. Externalized assumptions, more creative thinking, and informed decision-making.

In summary, anticipation can appear in design as both explicit and implicit affordances. A notable example is the widely discussed door handle by Don Norman in his book *The Design of Everyday Things* [66].

Design can trigger an anticipatory understanding of how the door operates (pull, push, slide) based on its visual cues. In the design field, anticipatory experiences involve using the willingness to foresee the future to facilitate decision-making, set future-oriented goals, and plan strategies for future design outcomes. Anticipation is an influential tool designers can use to shape the future by predicting potential user needs, behaviors, and desires.

The anticipation theory can improve anticipatory experiences in a wide range of circumstances. Although anticipation, anticipatory behavior, and anticipatory systems are often used indiscriminately, they all have distinctive characteristics that differentiate them from predictive systems:

- **Anticipation** is an overall capacity, not exclusive to humans, to act in response to preparation for a potential future reality [67].
- **Anticipatory behavior** is a process that depends on past and present events as well as predictions, expectations, or beliefs about the future.
- An **anticipatory system** is a system whose current state is determined by a future state. "The cause lies in the future" [68].

Are you designing for what comes next?

- Anticipatory systems aren't just about prediction—they're about alignment with human purpose. As you reflect on the chapter, ask yourself:

- Are my designs helping users achieve **their future goals**, or just responding to present needs?
- Am I using data to **simplify decisions**, or overwhelming users with options?
- Does my system consider **context and intent**, or only past behavior?
- Am I designing systems that are **flexible enough to adapt to changing user habits and futures?**

In the end, anticipatory design challenges us to move beyond optimization. It asks us to design with care, courage, and conscience. As we continue to create systems that foresee user needs, we must also anticipate the impact of our actions on individuals, businesses, and the planet. Design is no longer just about solving problems but about shaping futures.

TL;DR

From reactive to anticipatory systems—Anticipatory design shifts digital systems from waiting for user input to proactively supporting user goals:

- It replaces $A \rightarrow B \rightarrow C$ workflows with streamlined $A \rightarrow C$ logic, removing unnecessary effort.
- These systems use behavioral, contextual, and real-time data to reduce decision fatigue and ease cognitive load.

Prediction versus foresight—There are two distinct capabilities powering anticipatory systems:

- **Prediction**: Estimates near-term actions using historical and contextual data.
- **Foresight**: Designs for long-term outcomes by anticipating how needs may evolve.

Together, they support timely, relevant, and forward-thinking experiences.

Forecasting and backcasting—Forecasting helps anticipate future behavior by identifying patterns in the past. While backcasting starts from desired future outcomes and works backward to guide present-day decisions.

These two methods are complementary:

- **Forecasting** = short-term prediction
- **Backcasting** = long-term alignment with user goals

Circularity and causality—Anticipatory systems must understand *why* behavior happens, not just what happens through:

- **Forward circularity**: Future expectations shape current actions.
- **Backward circularity**: Present behavior updates future predictions.

Causal reasoning and counterfactuals (e.g., "what if?" scenarios) enhance the system's ability to adapt.

From generation to agency—Many AI tools today are generative—reactive systems that wait for prompts, like ChatGPT or Midjourney.

But anticipatory systems move toward **agentic AI**:

- They act, adapt, and learn with minimal intervention.
- This shift marks the emergence of AI that collaborates, not just creates.

Designing anticipation—Anticipatory systems intervene in decision-making at three levels.

- **Simplifying selections**: Pre-filled defaults and intelligent suggestions (e.g., Spotify)
- **Choice editing**: Narrowing options using context (e.g., Easysize)
- **Eliminating decisions**: Automating low-stakes choices without removing agency (e.g., Qapital)

Strategic use of explicit anticipation—Designers can apply anticipation across three futures-facing strategies:

- **Optimization**: Improve performance based on past patterns.
- **Contingency**: Prepare for uncertainty and edge cases.
- **Novelty**: Imagine radically different futures and user needs.

These strategies are used together to balance stability with adaptability, helping designers align present actions with future intent.

A New Design Direction
- Anticipatory systems challenge us to think beyond convenience and personalization:
- Are we designing systems that empower users, or nudge them?
- Are we building for immediate engagement or long-term transformation?
- Anticipatory design isn't just about solving problems—it's about shaping futures, ethically and intentionally.

Making Sense of AI

12

Users' Mental Models and Trust Gaps

As AI becomes increasingly embedded in digital experiences, understanding how users mentally model these systems is critical for building trust, usability, and engagement. This chapter explores users' challenges when interacting with AI-powered systems, including gaps in mental models, transparency issues, and loss of autonomy. Drawing on behavior science and UX research, this chapter presents actionable strategies for designing AI systems that foster trust, clarify capabilities, preserve user agency, and align ethical responsibility with user expectations.

Users rely on mental models—internal representations of how systems work—to navigate digital experiences. Past interactions, expectations, and familiar design patterns shape these models. When systems behave in ways users anticipate, experiences feel intuitive and trustworthy.

AI-powered systems, however, introduce a new layer of complexity. Their probabilistic nature, dynamic behavior, and often opaque decision-making processes challenge users' existing mental models. Without clear cues or familiar patterns to guide them, users struggle to predict how AI will behave, leading to confusion, misplaced trust, or misuse. Poorly aligned mental models can result in serious consequences: users might overtrust flawed recommendations, underutilize helpful features, or disengage entirely when outcomes feel inexplicable.

To design effective, trustworthy AI systems, teams must intentionally guide how users form mental models. This means designing with transparency, fostering appropriate levels of user control, and aligning system capabilities with user expectations. Good

DOI: 10.1201/9781003642800-15

AI design doesn't just solve problems—it builds bridges between human intuition and machine logic.

Shaping users' mental models is not optional. It's foundational to earning trust, enabling effective interaction, and realizing the full potential of AI-powered experiences.

12.1 MENTAL MODELS IN AI: THE NEW FRONTIER OF TRUST

Users build mental models—internal frameworks that help them predict how a system behaves—through repeated interactions and prior experiences. In traditional digital interfaces, where cause and effect are often straightforward, these models work well: click a button, see a predictable result. AI systems, however, challenge this dynamic. Unlike static interfaces, AI-driven experiences are probabilistic, adaptive, and sometimes unpredictable. Users often expect deterministic outcomes, believing that a particular input will always yield the same output. When AI behaves differently, it can confuse or frustrate users who are trying to form coherent mental models.

For instance, an intelligent assistant recommends a book reading list based on past reading habits. The user expects every book suggested to match their expectations and likes. If the recommendation list misses the mark, the user may not blame the variability of the AI model—they may instead question the assistant's competence or usefulness. This misalignment between expectation and reality erodes trust over time.

Building trust in AI systems begins with designing for users' mental models. Interfaces must clearly communicate system capabilities and limitations, helping users form realistic expectations about what AI can (and cannot) deliver. However, we must remember that uncertainty is inevitable when users interact with AI-driven systems. Unlike traditional software, AI doesn't always behave in predictable, linear ways. To bridge this gap, designers must make AI's behavior visible, understandable, and trustworthy.

Transparency helps users form accurate mental models and prevents "automation surprise," where the system behaves in unexpected or confusing ways. Therefore, there are a few design practices we can follow to mitigate this:

- **From the start, communicate what the system can and cannot do**: Before they rely on it, help users understand where the AI performs well and where it might make mistakes.
- **Use visual cues, simple explanations, and progressive disclosure**: Don't overwhelm users all at once; offer layered, easy-to-digest explanations as users engage deeper.
- **Set realistic expectations**: Avoid overpromising or overselling AI performance. It's better for users to be pleasantly surprised than bitterly disappointed.

For example, when Google Photos automatically groups faces into albums, it proactively displays a message: "Groupings may not always be accurate."

This small, simple warning sets user expectations early, minimizing frustration if the AI makes mistakes and maintaining trust even when errors occur. Transparency is not optional in AI design. It is a core ingredient for building trust, managing risk, and helping users make informed decisions about when and how much they rely on intelligent systems.

Designing for transparency in isolated features, like facial grouping in Google Photos, is just one layer. However, as AI systems grow more autonomous, we must also consider how *entire systems* balance human control with machine decision-making.

The interplay between **autonomy**, **automation**, and **human intervention** is fundamental in designing AI systems that meet user needs and inspire confidence. These dimensions often overlap, necessitating a careful balance to ensure that systems remain transparent, accountable, and capable of delivering the benefits of advanced automation. As shown in Figure 12.1, this model illustrates how each of these elements—human involvement, system automation, and decision-making autonomy—interacts and shifts depending on the design approach. Understanding this dynamic is crucial for setting appropriate user expectations, ensuring explainability, and maintaining meaningful control in AI-powered experiences.

12.1.1 Understanding Mental Models Across Automation, Autonomy, and Human Intervention

As users interact with AI systems, their mental models shift depending on their degree of control, how decisions are made, and how the system behaves over time. These shifts

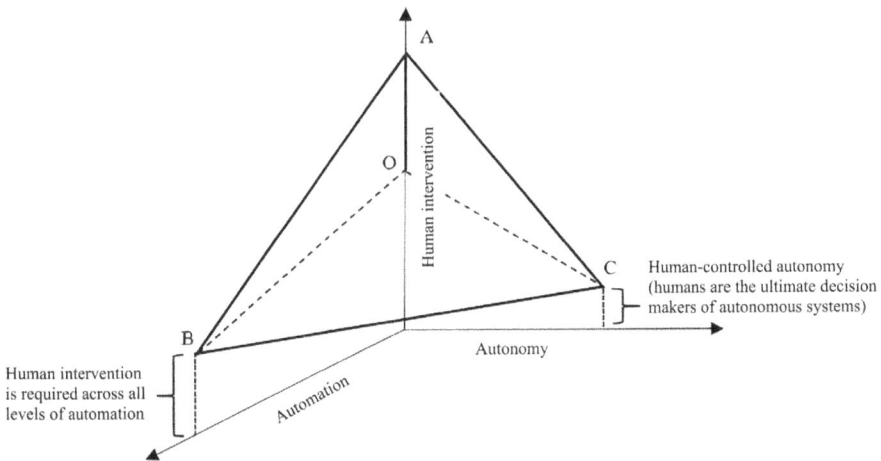

FIGURE 12.1   Interaction between autonomy, automation, and human intervention [36].

become particularly important when designing anticipatory systems, where the boundary between user and system agency is fluid. To design responsibly, we must consider how mental models vary across three intersecting dimensions: automation, autonomy, and human intervention. When users encounter **automated systems**, they tend to build relatively stable and straightforward mental models. These are based on predictability and repetition—click a button, get a consistent result. Automation gives users the sense that tasks are being handled *for them* within a framework they still feel they oversee. They expect rule-based logic and consistency. For example, when a thermostat adjusts the temperature automatically based on time of day, users accept this as a convenience. It offloads effort, but users still feel in control—they understand the rules driving the system.

In contrast, **autonomous systems** introduce unpredictability. Here, the system isn't just executing instructions; it's making decisions based on probabilistic logic, learning patterns, or optimizing for goals users may not fully understand. As a result, users often struggle to form coherent mental models. They may expect deterministic outcomes (if X does Y, Z will happen), but autonomous systems don't always follow this logic. This mismatch creates friction. A self-driving car, for instance, might decide to change lanes when the user least expects it. Even if the decision is correct, the user's trust may erode simply because the rationale wasn't clear or predictable.

Finally, users retain a stronger sense of control in systems that incorporate **human intervention**. The system supports their decision-making rather than replacing it. Here, mental models are shaped by collaboration: the user expects to guide the system, provide input, and even override its suggestions. These models tend to foster healthier trust calibration. Users understand that while the system may offer recommendations or automate certain tasks, they remain the final decision-makers. This balance preserves agency while still delivering the benefits of intelligent assistance.

In practice, few AI systems exist purely in one of these categories. Instead, anticipatory systems often blend these dimensions, requiring a careful balance to avoid undermining trust. Designers must recognize how system behavior—whether automated, autonomous, or collaborative—influences user expectations, perceptions of control, and willingness to engage. When designing across these boundaries, the goal is not to eliminate uncertainty but to help users navigate it with confidence. By acknowledging how mental models vary with the system's level of decision-making autonomy, we can

TABLE 12.1   Summary of Mental Models Across Automation, Autonomy, and Human Intervention.

DIMENSION	USER ROLE	MENTAL MODEL	TRUST RISK	DESIGN PRIORITY
Automation	Observer	"It runs the task for me."	Low (if predictable)	Reliability and consistency
Autonomy	Delegator	"It thinks and acts for me."	High (opaque logic)	Explainability and boundaries
Human Intervention	Decision-maker	"I steer; the system helps me."	Balanced	Control and feedback loops

better align our interfaces with user expectations and design experiences that are not just intelligent, but intelligible.

12.1.2 From System Structure to User Perception

While the structure of an AI system—whether automated, autonomous, or collaborative—plays a crucial role in shaping how it behaves, what truly matters for trust and usability is how users *perceive* that behavior. It's not just about the system's actual logic or capabilities, but about the *mental models* users form to make sense of those capabilities. These models are fluid, informed by interface signals, past experiences, and feedback loops—whether the user is an observer, a delegator, or a decision-maker.

But How Do Users Construct These Mental Models in Practice?
Cognitive science and UX research suggest three distinct outcomes that typically emerge when people interact with intelligent systems [69]. These outcomes reflect the user's understanding across different dimensions of interaction: how the system works, when it is appropriate to use, and what limitations it may have. Each of these is influenced by different design levers—value, control, and perceived transparency.

Mental models can be conscious or unconscious and can vary in complexity depending on an individual's level of expertise, familiarity with the subject matter, and ability to process and organize information. Mental models constantly evolve, and individuals may modify or adjust their mental models based on new information or experiences. Mental models can be applied to various domains, including decision-making, problem-solving, and learning, and can significantly impact an individual's behavior and performance.

These mental models include three outcomes:

1. **How does the system work?**—*How do you use it?* Users may develop an understanding of how the system works and how it arrives at its recommendations. The system's transparency influences this mental model.

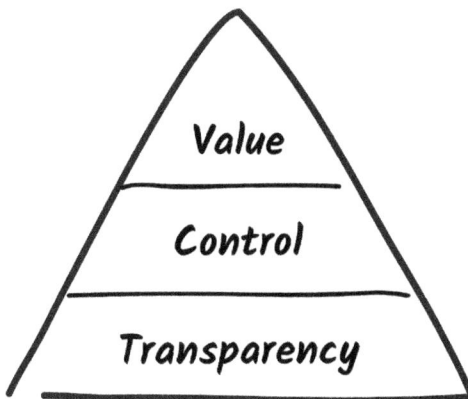

FIGURE 12.2   The interplay of trust: transparency, control, and value.

When to use it

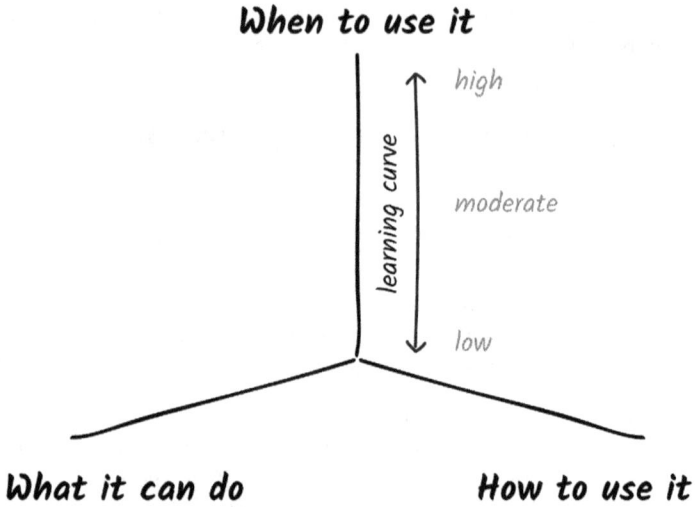

FIGURE 12.3   My interpretation of [69] user mental models' outcomes.

2. **System's impact on user work**—*When to use it?* Users may develop an understanding of how the system impacts their work and how they can best utilize the system to improve their productivity. The level of control users influences this mental model have over the system.

3. **System's limitations**—*What can it do?* Users may develop an understanding of the system's limitations, including when it may not be accurate or when it may not provide the best recommendations. The system's trustworthiness and value influence this mental model.

12.2 DESIGNING FOR TRUST: THE ROLE OF NOVELTY, AGENCY, AND CAPABILITY

Once users begin to form mental models of how an AI system behaves, shaped by transparency, control, and value, those models must be continuously supported by the experience itself. However, trust in AI systems is not monolithic. It is shaped by the context of use, the degree of user control, and the perceived value of the solution.

In particular, three interconnected dimensions play a critical role in shaping users' perceived trustworthiness: **novelty**, **agency**, and **capability**. Each reflects a different kind of psychological contract between the user and the system, and if not handled with care, each introduces unique risks.

- **Novelty** challenges users' understanding by placing them in unfamiliar situations. Without transparency, this can breed confusion or even mistrust.

- **Agency** touches on how much control users feel they have. Systems that override user input or behave opaquely can lead to feelings of powerlessness or surveillance.
- **Capability** addresses the question of whether AI is the *right tool for the job.* Users may disengage or push back if the technology feels excessive, misaligned, or ethically questionable.

Understanding and designing for these dimensions aren't just about improving usability—it's about building systems that users can *rely on.* The following sections explore how each dimension affects mental models, what design problems they introduce, and how to address them through intentional, human-centered strategies.

12.2.1 Novelty: Designing for First Encounters With AI

When users engage with an AI-powered system for the first time, they bring expectations shaped by past digital experiences, often rooted in deterministic, rule-based interfaces. But AI operates differently. It's probabilistic, dynamic, and, at times, unpredictable. In unfamiliar contexts, users instinctively lean on reliability and trust as coping mechanisms. Confidence falters when they don't understand how the system works—or why it behaves the way it does.

Why Transparency Matters?
In novel AI systems, transparency is not a luxury—it's a prerequisite for trust. Users must quickly understand what the system *can* do, *can't* do, and *might* do. Otherwise, they'll experience what researchers call "automation surprise": unexpected system behavior that feels arbitrary or inexplicable.

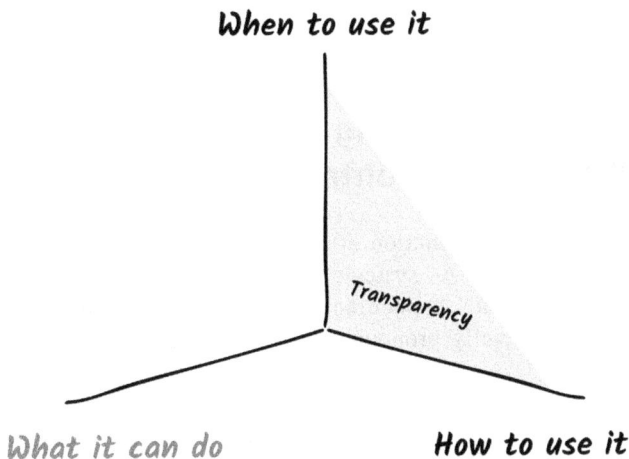

FIGURE 12.4 Transparency dimension under user AI mental models.

Making the inner workings of your AI system visible and understandable achieves several outcomes:

- Builds trust in the brand and system
- Prevents confusion and drop-offs
- Helps users access the full value of the experience
- Reduces cognitive load during unfamiliar interactions
- Supports smoother onboarding and easier long-term engagement

Designing for Trust and Transparency
Design teams should assist users in forming accurate mental models when deploying new AI systems. This involves the following actions:

- **XAI**: Provide clear, digestible explanations for key system behaviors or outputs.
- **Expectation management**: Tell users what the system can and cannot do before they rely on it.
- **Design for failure**: Assume the system will sometimes get it wrong. When it does, make recovery graceful, and take accountability.
- **Transparency of data use**: Be upfront about what data are collected and how they're used. Let users know they're working *for* them, not *on* them.

These principles aren't just ethical—they're practical. Transparency reduces friction, increases trust, and improves system adoption, especially when users are navigating AI for the first time. Ask yourself:

- **Explainability**: How will we help users understand unexpected outcomes?
- **Expectation management**: Are we overpromising? What happens when the system fails?
- **Accountability**: How will we handle errors, and how visible is our recovery process?

12.2.2 Agency: Balancing Automation With Human Control

In AI-powered systems, automation often promises convenience, but it can come at the cost of user agency. As systems take over more decisions, users may feel sidelined, unsure of what's happening or whether they're still in control. This loss of control is especially pronounced when AI behavior is opaque, complex, or inconsistent.

When to use it

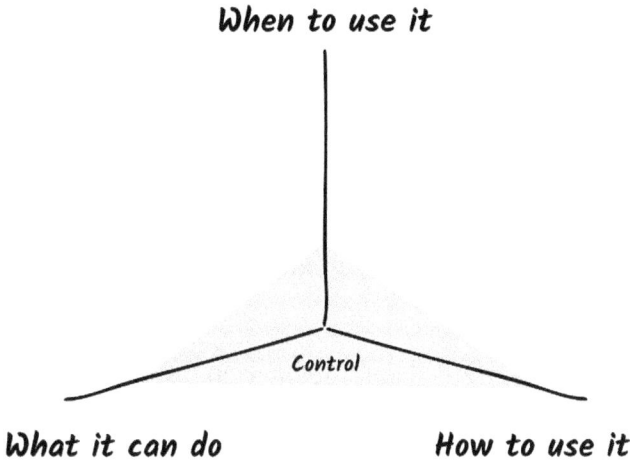

Control

What it can do   **How to use it**

FIGURE 12.5   Control dimension under user AI mental models.

Why Autonomy Matters?
Users need to feel *in charge*—not just of their data but of the decisions being made on their behalf. When AI systems operate too independently, people may question whether they're being watched, manipulated, or replaced. And once autonomy feels threatened, trust becomes difficult to rebuild. Designers must walk a fine line: offload cognitive effort without disempowering the user.

Designing for User Autonomy and Control
To support agency in high-automation environments, consider the following:

- **Offer meaningful controls**: Let users dial automation up or down based on context and preference.
- **Respect consent**: Don't assume users are always ready to hand over control. Let users choose to participate (opt-in) rather than making them disable something they didn't ask for (opt-out).
- **Support customization**: Let users shape the system's behavior, especially in contexts like personalization or recommendations.
- **Enable feedback**: Provide users with channels to correct, adjust, or retrain system behavior. This helps users *feel heard* and improves the model over time.

People are more likely to trust a system that collaborates with them, not one that tries to outsmart them. Ask yourself:

- **User feedback**—How will users shape or correct system behavior?
- **User autonomy**—Where can users fine-tune, override, or opt out?
- **Data privacy**—How clearly are we communicating how data is used, and why?

12.2.3 Capabilities: Aligning AI With Real Human Needs

Not every problem requires AI, and not every use of AI makes things better. In the race to adopt automation, it's easy to forget that technology should serve the problem, not the other way around.

Users build trust in a system not because it uses advanced tools but because it delivers meaningful and reliable value. They grow skeptical when that value is unclear or the system overpromises and underdelivers.

Why Value Alignment Matters?
Capability-based mental models form when users assess *what the system is actually good for.* The entire system feels overengineered if AI recommendations don't improve their decisions, reduce effort, or respect their context. Designers must ensure that AI's capabilities are aligned with:

- Real user problems
- Clear benefits
- Ethical and societal considerations

Designing for Value Alignment
To evaluate whether AI is truly adding value:

- **Start with the problem**: Ask "What's the best way to solve this?" before jumping to "Can we use AI here?"

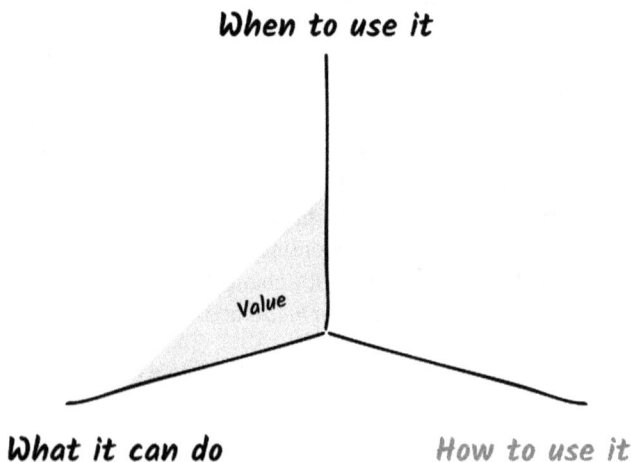

FIGURE 12.6 Value dimension under user AI mental models.

- **Prioritize the benefit, not the novelty**: Communicate how the system improves the user's life, not how it works behind the scenes.
- **Build for impact**: Consider not just usefulness but also fairness, inclusiveness, and long-term consequences.
- **Design for explainable outcomes**: Help users understand *why* a recommendation or decision was made.

Ask yourself:

- **Accountability**: Are we solving the right problem in the right way?
- **Fairness + inclusiveness**: Are we amplifying or reducing bias?
- **Ethics**: Do the benefits outweigh the potential harm?

12.2.3.1 Toward Trustworthy AI Experiences

As AI systems grow more autonomous and embedded in daily life, users are no longer just clicking buttons—they are forming beliefs, adapting expectations, and negotiating control in real time. Designers must recognize that mental models are not optional side effects—they are central to trust, understanding, and long-term engagement.

By addressing novelty, preserving agency, and grounding capabilities in real human needs, we can shape anticipatory systems that not only reduce friction but also respect the user's cognitive, emotional, and ethical boundaries. Trust isn't built on clever predictions but on clarity, alignment, and choice.

TL;DR

This chapter examines how users form mental models of AI-powered systems and why these models often break down due to AI's probabilistic, opaque, and adaptive nature. It explores how mismatches between user expectations and AI behavior can erode trust and usability, and how design can bridge this gap.

Mental models shape how users interpret, trust, and interact with AI. Poorly aligned models lead to confusion, misuse, and disengagement. Designers must actively support model formation to ensure that AI experiences remain intelligible, usable, and trustworthy.

Key insights:
- Users expect predictable outcomes from digital systems, but AI introduces unpredictability.
- Trust depends on transparency, meaningful user control, and the perceived value of the system.
- Novelty, agency, and capability are three critical factors influencing trust and must be addressed intentionally in design.

Design implications:
- Make system behaviors explainable and set realistic expectations.
- Allow users to calibrate automation levels and retain decision-making authority.
- Ensure that AI solves the right problem and that its use adds meaningful value without introducing ethical or social harm.

Trust in AI doesn't emerge from flawless performance—it emerges from transparent systems, respectful of agency, and grounded in real user needs. Designing for strong mental models is essential for enabling responsible and anticipatory AI experiences.

Designing Anticipatory Systems

A Layered and Temporal Approach

13

This chapter introduces a practical and structured approach to designing antici-patory systems. It examines how intelligent, future-facing experiences emerge from the interdependence of three core design layers: user intent as the strategic foundation, workflows that translate goals into processes, and algorithms as the adaptive execution layer. Each tier builds upon the last—intent provides direc-tion, workflows define the logic, and algorithms deliver responsiveness.

If anticipation is about designing for what comes next, then we need tools to help us see, shape, and plan for the future. This chapter introduces the **temporal mindset** necessary for anticipatory design—an approach that goes beyond moment-to-moment interaction to engage with long-term goals, evolving contexts, and future uncertainty. To design these systems well, we need to balance automation with human agency and understand how system autonomy impacts trust. This chapter introduces a layered approach—centered on **intent, workflows**, and **algorithms**—alongside temporal methods like **forecasting, backcasting**, and **retrospective thinking** to help create systems that are both proactive and aligned with user values.

These methods help bridge the gap between short-term prediction and long-term vision, ensuring that systems remain both technically robust and meaningfully aligned with human goals. In this way, this chapter connects anticipatory theory to actionable design, blending strategic foresight with real-world implementation.

13.1 FROM PERSONALIZATION TO ANTICIPATION

Designing for what comes next requires more than accurate algorithms—it demands a **temporal mindset**. Anticipation isn't just about responding to what users do in the moment; it's about aligning systems with where users are headed. That shift—from reactive assistance to proactive partnership—requires tools that help us **see, shape, and plan** for the future.

But understanding where anticipatory systems are going helps first to know where they've been. In their book *Human+Machine*, Daugherty and Wilson outline three historical waves of business transformation [70]. These waves offer a valuable lens for tracing the evolution of AI in enterprise contexts and help frame the current opportunity for anticipatory design.

Business-to-business (B2B) companies have long relied on AI-based solutions to automate routine operations and logistics tasks. The first wave of transformation focused on **standardizing processes**, a shift initiated by the first Industrial Revolution and carried forward through models like Fordism, Taylorism, and Toyotism. Daugherty and Wilson refer to this era as the *first wave* of business transformation.

The *second wave* emerged in the 1970s and reached its height in the 1990s, driven by **automated processes** and the rise of information technology. This era was defined by the business process reengineering movement and the widespread adoption of computers, large-scale databases, and automation tools. As back-office functions became digitized, many roles were displaced by machines capable of executing repetitive tasks more efficiently.

The *third wave* **centers on adaptive processes**, which the authors argue are just beginning to take shape. While it builds on the foundations of standardization and automation, it represents a fundamentally new way of operating that enables organizations to respond in real time to individual behaviors, preferences, and needs.

Once this third wave is fully achieved, both B2B and B2C businesses will harness AI, not just for increased efficiency but also for deep personalization. This advancement will allow them to deliver products and services that exceed what traditional mass production can offer, leading to greater agility and responsiveness value.

13.2 PERSONALIZED VERSUS TAILORED VERSUS ANTICIPATORY EXPERIENCES

As we move from standardized and automated systems into adaptive, real-time experiences, it's important to unpack the subtle but significant differences between *personalized*, *tailored*, and *anticipatory* experiences. These terms are often used interchangeably but operate at different levels of sophistication and intent.

Personalized Experiences
They are based on static user data—preferences explicitly shared or patterns observed over time. This might include saved settings, past purchases, or favorite categories. Personalization is reactive by nature: it adjusts once the system has gathered enough historical information.

Tailored Experiences
It takes things a step further by factoring in a real-time context. These systems adapt based on live inputs such as time of day, location, mood, or even biometric data. Tailoring is more dynamic than personalization but still relies on observable, surface-level signals.

Anticipatory Experiences
In contrast, anticipatory experiences seek to intervene *before* the user expresses a need. They rely on predictive modeling, behavioral analysis, and contextual foresight. These systems make decisions on the user's behalf—ideally minimizing effort and friction—without needing explicit input at the moment. Here's a quick comparison:

TABLE 13.1   Comparison Between Personalized, Tailored, and Anticipatory Experiences.

TYPE	DATA SOURCE	TIMING	EXAMPLE
Personalized	Historical preferences	Reactive	Netflix recommends based on watch history
Tailored	Real-time context	Adaptive	Spotify adjusts recommendations based on music listening history
Anticipatory	Predictive models	Proactive	Google Maps alerts you to traffic before you leave home

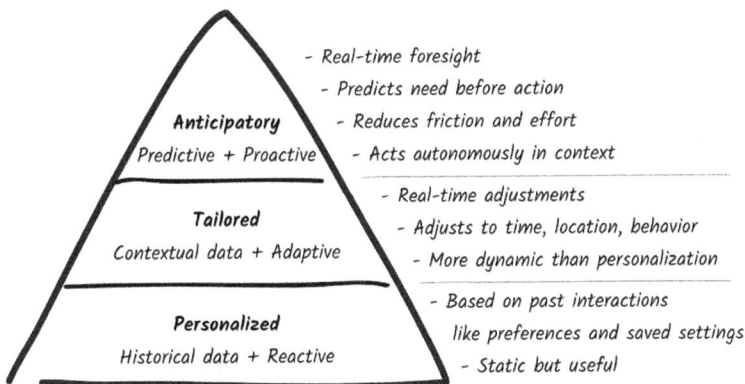

FIGURE 13.1   Levels of system proactivity.

Understanding these distinctions clarifies what it truly means to *design for intelligence*. As we move toward increasingly autonomous systems, we must carefully consider the trade-offs between effort reduction and user control. When does helpful become intrusive? When does automation become overreach?

This book explores those questions by framing anticipatory design as a discipline of both empathy and foresight—one that demands not just technical fluency but a deep understanding of what users need before they even know to ask.

13.3 THE THREE-LAYER MODEL OF ANTICIPATORY SYSTEMS

If we want to design systems that feel truly intelligent—not just responsive but meaningfully proactive—we need a clear structure to guide their construction. Anticipatory systems aren't magic. They're built from the inside out, starting with the user's *intent* at the core (what they genuinely want to achieve), surrounded by *workflows* that define the steps needed to reach that goal, and finally wrapped in *algorithms* that enable those steps through predictive intelligence.

This section introduces a three-layered model for designing anticipatory systems, offering a practical lens to connect theory and implementation.

This model—composed of **intent**, **workflows**, and **algorithms**—helps us zoom in on the different layers of intelligence that work together in any anticipatory product. Think of it like an onion: at the center is what the user wants to achieve; wrapped around that are the steps or logic that get them there; and finally, surrounding it all are the predictive technologies enabling those steps to unfold, often before the user even lifts a finger.

13.3.1 *Intent*: What Does the User Truly Want?

At the heart of any anticipatory system is **intent**—the user's underlying future goal. It's not always what they click on or say aloud, but what outcome do they genuinely want to achieve? For example, with a product like the Nest thermostat, the user's intent might be simple: stay comfortable while using as little energy as possible. That intent becomes the foundation for everything the system does next. When design begins from this point of view, systems don't just function—they feel aligned with what matters to users.

13.3.2 *Workflows*: Turning Intent Into Action

In the next layer, we have **workflows**, representing the steps a system uses to fulfill that intent. In Nest's case, these workflows might include learning a user's temperature

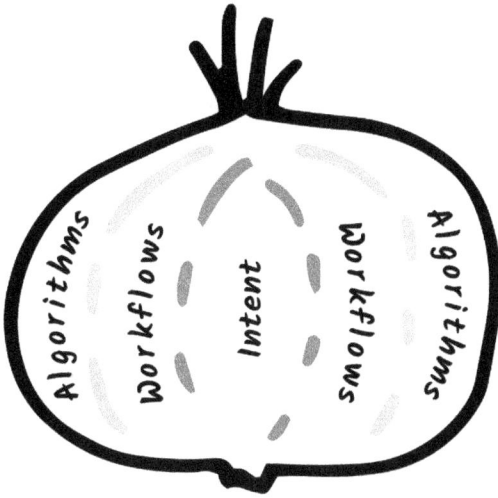

FIGURE 13.2 The architecture of intelligent systems, from intent to algorithms.

preferences over time, factoring in seasonal conditions, or adjusting the heating when nobody's home. These steps are structured, intelligent, and invisible to the user. But together, they form the bridge between raw data and meaningful action.

13.3.3 *Algorithms*: The Predictive Engine

Surrounding it all are the outermost layer—**algorithms**—which include DL models that spot patterns, make predictions, and decide how and when to act. In a platform like Thrive, for example, algorithms might analyze mood trends, sleep data, or engagement with stress-reducing content to deliver tailored recommendations. It's here that anticipatory systems show their full power—predicting needs before they arise, nudging users toward healthier habits, or optimizing a plan based on real-time behavior.

Let's break it down using Thrive:

- **The intent (core layer): Enhancing overall well-being** by assisting users to achieve a healthier and more fulfilling life. This includes stress management, improving sleep quality, and increasing energy levels.
- **Workflows (middle layer): Tailored programs and support utilizing user** data (sleep patterns, activity levels, mood) to create customized programs that meet their tailored needs and goals. These programs involve various workflows, including the following:
  - Guided meditations and breathing exercises to manage stress and anxiety.
  - Personalized sleep routines are designed to improve sleep quality.
  - Educational content and coaching tips to promote healthy habits and lifestyle changes.

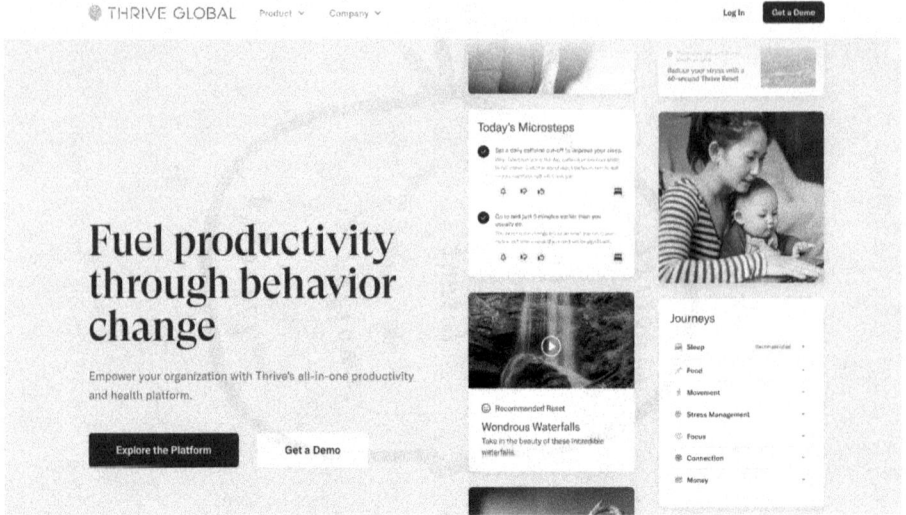

FIGURE 13.3   Screenshot of the Thrive website, a personalized well-being platform, as it appears in 2025.

Source: Accessible at thriveglobal.com

- **Algorithms (outer layer)**: Using algorithms to analyze user data, these systems provide **data analysis and personalized recommendations** to generate actionable insights. These algorithms perform tasks such as:
  - Identify patterns in sleep, activity, and mood to understand user challenges.
  - Predict user behavior to recommend interventions that address potential issues.
  - Optimize program recommendations based on user progress and data analysis.

By aligning algorithms and workflows with the core intent of improving well-being, Thrive provides a personalized and proactive approach to behavior change. Here's how it benefits users:

- **Sustained behavior change**: Personalized programs and ongoing support empower users to develop healthy habits for the long term.
- **Data-driven insights**: User data analysis helps users gain valuable insights into their well-being and identify areas for improvement.
- **Proactive support**: Anticipates potential issues and recommends interventions before problems arise.

Thrive exemplifies how anticipatory design can effectively achieve positive life outcomes, aligning seamlessly with this layered approach.

13.4 WHERE TEMPORAL METHODS LIVE IN THE MODEL?

This layered approach clearly demonstrates how tailored, proactive support is crafted. The system doesn't wait—it meets users where they are and gently moves them forward.

The future-oriented nature of anticipation makes it both appealing and challenging to implement. Anticipation's future-facing nature makes it both attractive and complex to implement. To overcome these challenges, you can leverage two complementary approaches: **forecasting** and **backcasting**.

- **Anticipation** involves making explicit predictions about future states or events, which is achieved through the following:
    - Forecasting involves evaluating historical trends and data to predict future outcomes. By examining previous data and current behaviors, designers can foresee user journeys that explore future needs. This method supports strategic planning and decision-making based on projected future scenarios.
    - Backcasting, conversely, starts with defining a desired future outcome and working backward to identify the steps needed to achieve that goal. This approach ensures that the service is aligned with long-term user goals and visions rather than merely responding to immediate needs.
    - Foresight is a systematic approach to analyzing and anticipating future changes in a wide range of fields, including technology, politics, economics, environment, society, and culture. It includes considering alternative future scenarios and making informed decisions to navigate toward the desired outcomes.

Intent onion diagram

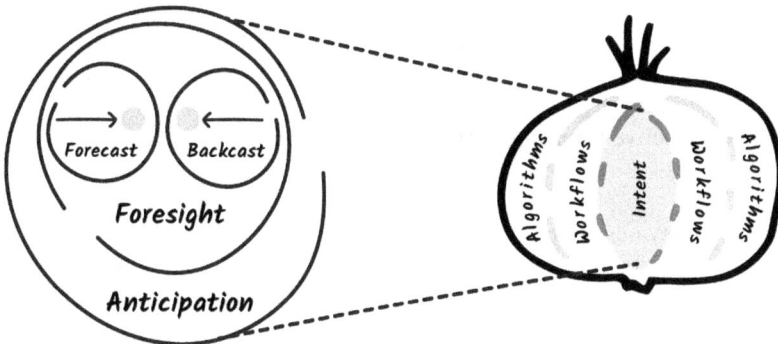

FIGURE 13.4 The role of anticipation and foresight in aligning system design with user intent.

Remember: One of the hardest parts of anticipatory design is designing for a future that hasn't happened yet. Therefore, it is essential to incorporate forecasting and backcasting into the process for the following reasons:

- **Forecasting** uses historical and current data to predict what will likely happen next.
- **Backcasting** starts with a future goal and works backward to determine the steps to get there.

Both methods help systems stay grounded in long-term thinking rather than short-term responses, especially when combined with foresight and the systematic exploration of future possibilities. With the proposed layer approach, complemented by anticipation and foresight, you can design anticipatory systems that provide tailored and proactive support to your users. By integrating forecasting, backcasting, and considering alternative future scenarios, these systems help users achieve significant life outcomes instead of merely fulfilling isolated tasks.

Building on the layered structure of intent, workflows, and algorithms, anticipatory systems must incorporate additional dimensions to maximize their efficacy:

- **Feedback mechanisms**: Learning from how users interact with the system, like when someone overrides a thermostat setting or skips a wellness suggestion.
- **Transparency**: Letting users see what's happening and why, especially in high-stakes environments like health or finance.
- **Context awareness**: Pulling in external data—weather, schedules, and environment—to improve relevance.
- **Adaptability**: Evolving over time as users grow, learn, and change their routines or goals.

Foresight acts as an umbrella term for **temporal design methods** that assist us in exploring and shaping potential futures. Typically, this includes the following items:

- **Forecasting**: Projecting trends based on historical and current data
- **Backcasting**: Starting from a desired future and working backward to identify the necessary steps
- **Retrospective thinking**: Drawing insights from past events to inform future strategies

As discussed previously, prediction estimates what might happen, while anticipation helps shape what should happen. Anticipatory systems rely on both using predictive data to act in the moment and foresight techniques to align those actions with meaningful long-term outcomes.

Prediction is understood as an assessment of uncertainty involving methods like forecasting (predicting future states), backcasting (goal-oriented planning), and retrospection (examining past events to guide future planning) to shape desirable futures that can be integrated into design practice [30, 71, 72]. Together, they enable designers to model uncertainty, explore possible scenarios, and build systems that don't just react but guide and adapt to what's next.

This indicates that anticipation encompasses more than just predicting the future. These methods facilitate temporal analysis (looking at different points in time) and projection (predicting or planning future outcomes), which is essential for supporting the anticipatory design process [31]. These techniques support the anticipatory design process by allowing designers to model how different futures unfold.

13.5 USE CASES OF TEMPORAL METHODS

To bring these concepts to life, let's examine how temporal methods like forecasting, backcasting, and retrospective thinking are being integrated into real-world anticipatory systems—not only to improve usability but also to support long-term behavior change, decision-making, and innovation. Each method contributes a distinct temporal lens to anticipatory design:

- **Forecasting** excels at identifying trends and making preemptive recommendations based on historical patterns.
- **Backcasting** anchors system design in intentional futures, helping align workflows with user-defined goals.
- **Retrospective thinking** supports creative exploration and risk mitigation by reimagining the past from a future vantage point.

By combining these lenses, designers can better calibrate user experience for uncertainty, ensuring that anticipatory systems don't just automate or predict but empower [30, 72].

These approaches are not mutually exclusive. In practice, the most robust anticipatory systems integrate multiple temporal strategies. For instance, a health platform might use the following:

- **Forecasting** to anticipate user stress based on biometric and contextual data.
- **Backcasting** to map the user's long-term goal of improved sleep or energy levels.
- **Retrospective thinking** to simulate outcomes and fine-tune interventions based on hypothetical success scenarios.

Together, these methods extend the designer's toolkit beyond user-centered thinking to future-centered thinking. They help ensure that anticipatory systems remain responsive in the short term while remaining aligned with deeper human values, long-term intent, and the shifting conditions of the world around them.

13.6 IMPLEMENTING FORECASTING IN ANTICIPATORY SYSTEMS

Forecasting is one of the most commonly used temporal methods in anticipatory design. At its core, it relies on historical data and patterns to predict what might happen next [71]. This process involves selecting pertinent data points, rigorously evaluating variables, and formulating predictions or calculations [72]. Designers and developers use forecasting to make smarter, more proactive decisions—ideally before users even realize they need help.

One compelling real-world example is **Salesforce Einstein Analytics**, a cloud-based business intelligence tool. Einstein analyzes historical trends, customer behaviors, and key business metrics to deliver automated insights and predictions. This allows businesses to act in advance—whether that's anticipating sales dips, suggesting product improvements, or streamlining customer interactions.

To break it down, forecasting typically involves three core steps:

1. **Data selection**: Identifying the correct historical data that can inform predictions
2. **Variable evaluation**: Analyzing the factors that may influence future outcomes
3. **Outcome prediction**: Using the data models to forecast potential future scenarios

This structured process can be incredibly helpful for proactive planning, especially when teams must identify signals pointing toward possible opportunities or risks. But forecasting isn't foolproof—and that's important for everyone to remember.

Strengths of Forecasting
- **It builds on success**: Forecasting leverages past achievements (what works) to project future possibilities. This data-driven approach helps identify trends and anticipate potential outcomes.

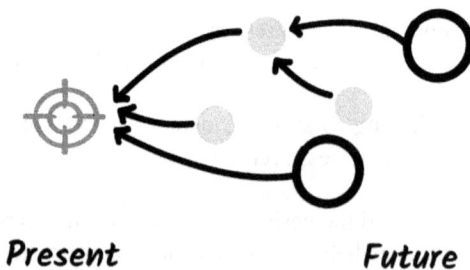

Present   Future

FIGURE 13.5   Contrasting forecasting and backcasting approaches.

FIGURE 13.6   Illustration of Salesforce Einstein AI service [73].

- **It deepens understanding**: Analyzing what's worked (and what hasn't) in the past can clarify the root causes of current challenges.

Limitations of Forecasting
- **Optimism bias**: Forecasting can be overly optimistic by focusing on past successes. This "extrapolation bias" can lead to inaccurate predictions, overlooking potential challenges.
- **Complexity challenges**: Forecasting may not be sufficient for planning in complex, dynamic systems. These systems involve numerous interacting factors that are difficult to predict using solely historical data.
- **Data dependence**: Forecasting is heavily reliant on data quality. Biased datasets can skew the results, leading to inaccurate or misleading predictions.
- **Big data's double-edged sword**: The rise of big data presents both opportunities and challenges for forecasting. While vast amounts of data can provide more comprehensive insights, the potential for bias also increases.

In essence, forecasting involves extrapolating current data and trends to envision future possibilities and explore the long-term outcomes technology can enable. The extrapolation process uses historical patterns to predict what might come next. However, relying heavily on this method can introduce what's called **extrapolation bias**—the tendency to assume that recent trends will continue indefinitely. For example, if a company has grown its earnings by 25% annually over the past three years, it's tempting to expect that growth rate to persist far into the future is tempting. But this kind of projection often ignores other important variables and becomes increasingly inaccurate the farther out we forecast. In addition, forecasting alone may fall short in complex systems where many interconnected and dynamic elements interact unpredictably. These intricate systems cannot be accurately modeled using past data alone. Another limitation lies in the quality of the data itself. Biased or incomplete datasets can skew forecasting results, leading to inaccurate or even harmful predictions. And while the rise of big data offers richer information for forecasting, it also amplifies the risk of hidden bias within those massive datasets [72].

13.7 IMPLEMENTING BACKCASTING IN ANTICIPATORY SYSTEMS

Where forecasting looks at where trends are heading, **backcasting** takes a different approach. Instead of using data to predict the future, it starts with defining a desired future outcome—a clear user intent—and works backward to determine the necessary steps to achieve it [71]. The process then works backward to identify the steps needed to achieve that goal.

This goal-oriented approach crafts an experience that actively guides users toward their desired future state. For instance, a financial planning app might start with a user's long-term financial goal, such as retirement savings, and then design an experience that guides the user through each step necessary to reach that goal, from budgeting tips to investment recommendations. Backcasting is often considered more intuitive and suggestive because it is driven by a specific goal [74]. It focuses on:

- Recognizing patterns and connections between seemingly unrelated events
- Exploring potential causal chains that could lead to a desired outcome
- Developing metaphors or theories to inform future-oriented planning

A prominent example of backcasting in action is You Need a Budget (YNAB), a financial planning app designed to help users achieve their savings and budgeting goals.

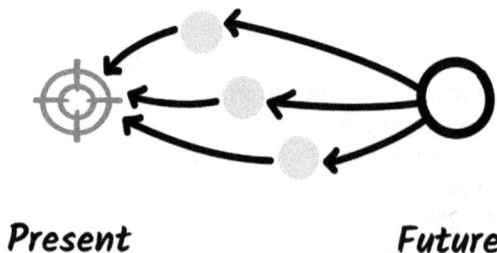

Present   Future

FIGURE 13.7   Understanding backcasting.

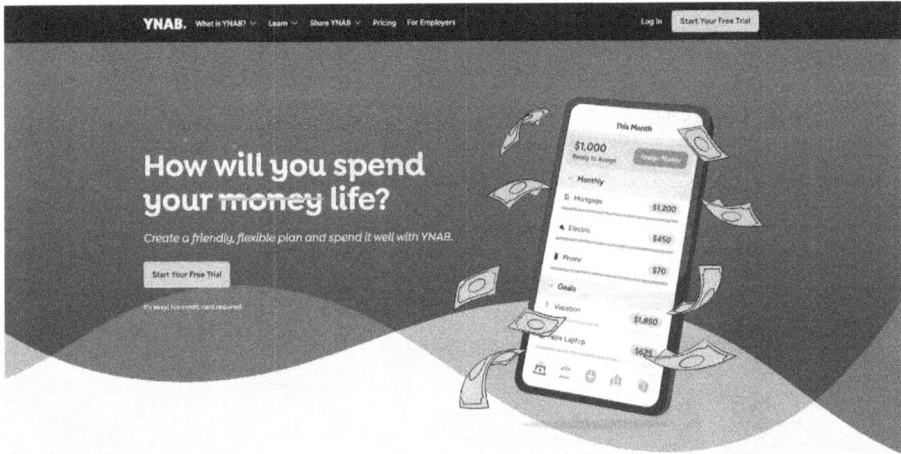

FIGURE 13.8   A screenshot of the YNAB website, as it appears in 2025.

Source: Accessible at ynab.com

YNAB starts by asking users to define their financial targets, such as saving for an emergency fund or paying off debt within a specific timeframe. The app then works backward to outline a step-by-step plan, dynamically adapting to the user's financial behavior. By analyzing spending patterns, income, and historical trends, YNAB provides actionable recommendations, such as reallocating funds between categories or reducing discretionary expenses. As users deviate from or exceed the plan, the system recalibrates to keep progress aligned with the target. YNAB exemplifies backcasting effectiveness by aligning financial tools with user intent and empowering individuals to achieve long-term goals.

Similarly, FitnessAI showcases the effectiveness of backcasting by designing adaptive fitness plans aligned with user-defined goals, dynamically adjusting based on progress. This mobile app leverages AI to help users plan their fitness journeys. Instead of simply tracking workouts, FitnessAI allows users to set a desired future fitness state (e.g., run a marathon). The app then uses backcasting principles to create a personalized workout plan, including recommended exercises, sets, reps, weights, and rest periods. FitnessAI continuously tracks progress and adapts the plan in real time, ensuring that it remains aligned with the user's goals. This personalized approach effectively utilizes backcasting to guide users toward their desired fitness outcomes.

Backcasting plays a vital role in shaping the direction of technological development. By starting with a vision for a sustainable future or human behavior, backcasting can identify the specific technological advancements needed to achieve that vision. This targeted approach helps researchers and developers focus their efforts on creating solutions that address sustainability challenges.

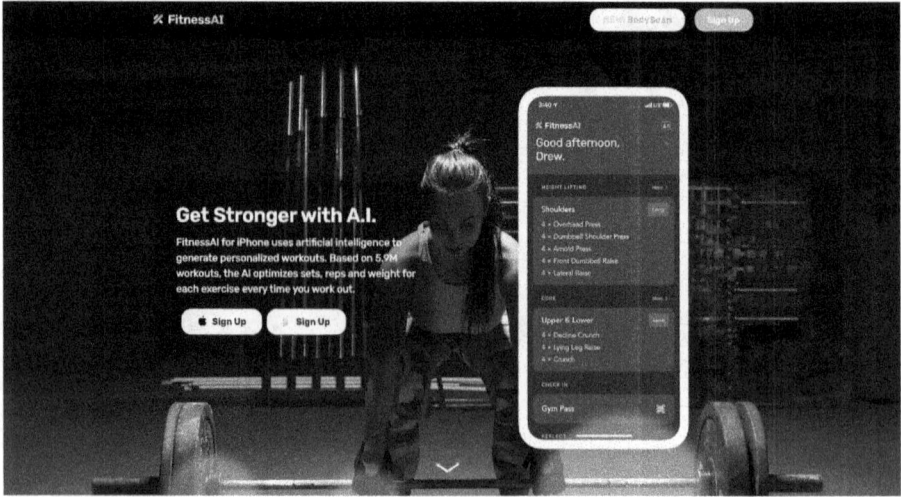

FIGURE 13.9   A screenshot of the FitnessAI appçication website, as it appears in 2025.

Source: Accessible at fitnessai.com

13.8 IMPLEMENTING RETROSPECTIVE THINKING IN ANTICIPATORY SYSTEMS

Retrospective thinking, often associated with recalling past events, can be a powerful tool in anticipatory design [72]. This approach involves imagining a hypothetical future event as if it has already happened and then working backward to understand the steps that led to it. By analyzing this fictional past, designers can identify potential unknown system states and the paths that could have led to them [71]. Therefore, retrospective thinking is particularly valuable in innovation-focused projects. For instance, imagine the future of AI services with generative AI. Picture a design team developing a groundbreaking AI service for personalized education. Using retrospective thinking, they might project themselves five years into the future, celebrating their AI tutor's widespread adoption and success. By working backward through this hypothetical past, the team can identify critical factors contributing to its success.

A generative AI tool, like Gemini or ChatGPT, could be particularly valuable. The team could prompt the AI with details about their envisioned future success, such as specific awards or positive user reviews. The generative AI could then create a fictional narrative describing the journey that led to this success. This narrative might highlight unforeseen challenges the team overcame. User needs they hadn't initially anticipated, or pivots they made in their design approach. By analyzing this AI-generated fictional past, the design team can gain valuable insights and identify potential blind spots in their current plans.

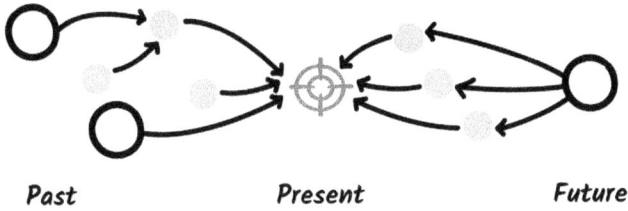

FIGURE 13.10   Understanding retrospective thinking.

FIGURE 13.11   ChatGPT on the left and Google Generative AI Tool on the right.

Designers naturally shift between temporal perspectives throughout the creative process, moving from the future to the past and back again [72]. We can further break down anticipatory thinking into three core processes, regardless of the specific approach (forecasting, backcasting, or retrospective) [71]:

- **Situational recognition**: Drawing on past experiences to interpret current cues and understand the situation.
- **System state extrapolation**: Envisioning how the system might evolve from its current state to a different state.
- **Mental model construction**: Developing a mental representation of the system based on available evidence, even if that evidence is incomplete.

These temporal methodologies emphasize that successful anticipatory systems are not merely reactive but are equipped to envision, plan for, and actively shape desirable futures.

13.8.1 Why It Matters for Design?

Anticipatory systems should be more than smart—they should be supportive. When we design with intent at the center, build thoughtful workflows, and power them with predictive intelligence, we move closer to technology that empowers rather than overwhelms.

When we combine this model with strategic foresight tools, we gain the ability not just to respond to the future but to shape it.

TL;DR

The layered model: Designing effective anticipatory systems requires aligning three layers—the **intent** (what users want), **workflows** (how the system helps), and **algorithms** (what powers it).

Temporal design methods:
- **Forecasting**: Uses past data to predict future outcomes.
- **Backcasting**: Starts from a desired future and plans backward.
- **Retrospective thinking**: Imagines the past from a future point to uncover blind spots and opportunities.

Why it matters? Anticipatory design isn't just technical—it's ethical, strategic, and future-facing. Done right, it supports users meaningfully, adapting to them rather than merely reacting to them.

Designing With Foresight

14

Future Studies in Anticipatory Design

Designing anticipatory systems demands more than reacting to current patterns—it requires shaping experiences that evolve with users over time. This chapter introduces strategic foresight, a discipline rooted in future studies, and explores how it can complement temporal methods like forecasting and backcasting. We examine why foresight skills are essential for designers today, how anticipation differs from prediction, and how to practically integrate futures thinking into UX, product, and service design. By adopting foresight, designers can create systems that are not only adaptive and resilient but ethically aligned with human-centered futures.

Anticipatory design transcends merely reacting to current trends; it requires a strategic approach that encompasses foresight and long-term vision. To create systems that are not only proactive and intentional but also inherently resilient, designers must engage in a deliberate process that considers future implications and users' evolving needs.

This is where foresight comes in. Rooted in the discipline of future studies, foresight is the practice of exploring multiple plausible futures to inform better decisions in the present. Unlike forecasting, which projects forward from historical data, foresight invites designers to explore uncertainty, imagine preferable futures, and shape present-day actions with long-term impact in mind. This chapter explores how strategic foresight and its associated tools—such as scenario planning, trend analysis, and environmental scanning—can be integrated into anticipatory design. It complements temporal methods like forecasting and backcasting, while also introducing anticipation as a distinct, present-oriented way of designing with futures in mind.

This section explores how foresight can complement temporal methods like forecasting and backcasting in anticipatory design.

DOI: 10.1201/9781003642800-17

14.1 WHY FORESIGHT BELONGS IN ANTICIPATORY DESIGN?

If we want to design truly anticipatory systems—not just predictive—we must move beyond forecasting alone. Forecasting excels at projecting trends from past behavior. Still, as we've seen throughout this book, it often struggles when users evolve unexpectedly, contexts shift rapidly, or edge cases emerge. The failure of early anticipatory systems like **Digit** was not due to a lack of data or analytics. It was a failure of imagination: the inability to foresee user journeys that deviated from the "average" path.

Anticipatory design asks us to build for what users will need, not just for what they need right now. However, real user needs do not move in straight lines. They grow, diverge, surprise, and sometimes contradict the models we build for them. Without foresight, anticipatory systems risk becoming brittle—optimized for yesterday's patterns rather than tomorrow's realities.

Foresight belongs in anticipatory design because it builds resilience, not just efficiency. Forecasting, based on patterns and probabilities, can tell us what is *likely* to happen next. But foresight challenges us to confront a more complicated truth: the future is not a single path—it's a branching network of possibilities. If designers only prepare for the "most probable" future, they leave systems fragile in the face of disruption, blind to outliers, and poorly equipped to support evolving human needs.

In a world defined by complexity and volatility, the cost of ignoring foresight is growing. **Forecasting predicts. Foresight prepares.**

14.1.1 Why Designers Must Learn Foresight Skills?

The evolution toward anticipatory systems demands that designers think beyond immediate user needs and project scenarios into dynamic, uncertain futures.

This means developing a new set of critical skills:

- **Scenario thinking**—Not assuming one future, but actively designing for multiple, branching possibilities.
- **Strategic flexibility**—Building adaptive workflows that allow systems to flex when new realities emerge.
- **Signal interpretation**—Reading weak signals, early indicators of change, to anticipate disruptions before they fully surface.
- **Ethical foresight**—Considering the second- and third-order consequences of anticipatory systems on users, society, and the environment.

Without foresight-driven adaptability, anticipatory systems risk becoming obsolete—optimized for past behaviors rather than future realities. Over time, this disconnect erodes user trust, diminishes autonomy, and undermines the very promise of anticipatory design: reducing friction while empowering meaningful action.

When anticipatory systems rely solely on historical data, they may initially succeed in minimizing effort by automating frequent decisions. However, without foresight-driven adaptability, these systems risk disconnecting users from important choices, diminishing user agency, and obscuring the real-world consequences of their actions. Over time, this can lead to user confusion, erosion of trust, and reduced satisfaction, especially in high-stakes contexts such as financial management or healthcare.

Ultimately, anticipatory systems that fail to incorporate foresight are vulnerable to becoming obsolete. They risk offering users outdated or inappropriate support, introducing new usability challenges instead of resolving them. To remain effective, anticipatory systems must anticipate not only what users need now but also how user needs, goals, and environments are likely to evolve. Therefore, designers who embrace foresight will not only design better systems but also design systems that survive longer, adapt better, and serve users more ethically.

14.1.2 Why Foresight Has Been Missing From Design Education?

Despite its growing relevance, foresight methodologies have historically been underrepresented in mainstream design curricula. This gap stems from several systemic factors rooted in the origins and evolution of the design discipline. First, early design education was shaped by industrial models of problem-solving. Traditional design frameworks, such as the design thinking process and the double diamond model developed by the UK Design Council, emphasize a linear, step-by-step approach: first define the problem, then explore solutions, then converge on an outcome. These models were optimized for relatively stable environments where user needs could be reliably uncovered and addressed within the bounds of a project cycle. They implicitly assumed that problem spaces were understandable, bounded, and slow to change. As a result, design education focused primarily on producing immediate, user-centered improvements rather than preparing for systemic change, disruption, or future uncertainty. Second, the structure of business incentives reinforced short-term thinking. Design work became tightly aligned with product development timelines, marketing cycles, and quarterly financial goals. Metrics such as time-to-market, conversion rates, and engagement statistics rewarded short-term optimization over long-term resilience. Within these constraints, success was measured by immediate outputs rather than by the system's adaptability to future shifts or emerging risks. Third, foresight requires interdisciplinary competencies that are not traditionally integrated into design education. Strategic foresight draws upon disciplines such as systems thinking, economics, and futures studies—areas that extend beyond conventional human-centered research methods. Developing comfort with uncertainty, complexity, and long-term planning requires a broader educational foundation than what most design curricula have historically provided.

14.1.3 The Shift Designers Must Now Make

The impact of foresight is inherently difficult to measure in the short term. While usability improvements and incremental optimizations yield fast, tangible results, foresight outcomes, such as building system resilience, mitigating emergent risks, or enabling future readiness, are often visible only over extended time horizons. This made foresight seem less immediately "actionable" compared to other design deliverables, and thus, it was deprioritized.

In short, given the accelerating evolution of technology and the growing volatility of the environments we design for, the traditional foundations of design education—built around stability, predictability, and known parameters—are no longer sufficient. Foresight is now a core professional competency, critical for designing resilient, adaptive, and ethically aligned systems over time.

14.2 FROM FORECASTING TO FORESIGHT: WHAT'S THE DIFFERENCE?

In user experience design, understanding how users behave today is essential—but anticipating how users might behave tomorrow is becoming equally critical. As businesses move toward anticipatory systems, it's important to distinguish between **forecasting** and **foresight**, two methods often confused but fundamentally different.

14.2.1 Forecasting: Projecting Probabilities

Forecasting predicts future events based on historical data and existing trends. It is quantitative, focused, and probabilistic. A forecast usually addresses questions such as:

- How many users will adopt a new feature based on last year's data?
- What are the chances that a user will churn given current behavior patterns?

Forecasting assumes that **past behaviors will largely predict future outcomes**—a useful but inherently limited approach. While forecasting is highly valuable for short-term planning, it often struggles to account for discontinuities, disruptions, or emerging user needs.

As discussed earlier in Chapter 7, several early anticipatory services—such as **Digit**, **LifeBEAM Vi**, and **Mint**—illustrate this challenge. These systems were built on the assumption that historical user data would remain stable and predictive. However, when user behaviors shifted due to new financial habits, changing health priorities, or unexpected life events, the systems could not adapt. As a result, what initially felt seamless became irrelevant or even frustrating to users. Without the foresight to imagine alternative futures, even the most sophisticated forecasting models risk leaving systems vulnerable to real-world complexity.

14.2.2 Foresight: Exploring Possibilities

Foresight, on the other hand, is **qualitative, exploratory, and possibility-driven**. Rather than projecting one likely outcome, foresight helps designers and stakeholders imagine a **range of plausible, possible, and preferable futures**—including futures where today's assumptions may no longer hold.

Foresight acknowledges that:

- The future is **not predetermined**.
- User behavior, societal shifts, and technological innovation can **diverge sharply** from historical patterns.
- Preparing for multiple futures is more resilient than optimizing for a single expected one.

For example, as discussed in Chapter 7, Mint emerged as a pioneer in personal finance applications by analyzing user behaviors. It compiled spending data, classified expenses, and estimated future cash flow based on historical trends. This forecasting approach worked—until it didn't. As new financial behaviors emerged (e.g., economic inflation, market crashes, cryptocurrency booming, subscription-based spending models), Mint's static categories and assumptions became increasingly misaligned with real user needs.

If Mint had applied foresight, it could have explored multiple emerging financial scenarios rather than relying solely on historical banking data. For example, using **scenario planning**, Mint might have envisioned futures where:

- Users no longer receive steady paychecks but manage multiple active and passive income streams.
- Subscription services (Netflix, Spotify, Patreon) dominate monthly budgets.
- Digital assets like crypto or peer-to-peer lending reshape savings behaviors.

Based on these foresight insights, Mint could have proactively

- Developed budgeting workflows that flexibly adapt to volatile, irregular incomes.
- Helped users plan for subscription fatigue and automate cancellation reminders.
- Integrated new asset classes like cryptocurrencies into wealth management tools.

In short, forecasting would have kept Mint optimizing for yesterday's salary-and-savings users; foresight could have helped it evolve with new financial realities, keeping users engaged, supported, and trusting the system through change.

14.2.3　Forecasting Versus Foresight: A Visual Summary

Let's return to our Matryoshka dolls to understand the relationship between forecasting, foresight, and future studies. Understanding this relationship helps clarify how different future-oriented methods support anticipatory design.

Anticipatory design, as a future-facing practice, demands tools that reach beyond immediate prediction. It requires methods that help us imagine broader, longer-term possibilities. Foresight, rooted in the field of future studies, offers exactly this: a strategic approach to exploring multiple futures and informing design decisions under uncertainty. Unlike forecasting, which concentrates on quantifiable trends and aims to predict likely outcomes, foresight embraces complexity. It is a qualitative, holistic process that helps designers prepare for a range of plausible, possible, and preferable scenarios, not just the most probable ones. In short:

- **Forecasting** narrows future thinking to what is likely.
- **Foresight** broadens future thinking to explore what could happen—and what should happen.

Future studies, where foresight has its roots, emerged formally in the 1950s as a military planning tool. By the 1960s, it entered the business world, evolving into a broader discipline focused on managing long-term change. As the United Nations Development Programme notes, "The premise of foresight is that the future is still in the making and can be actively influenced or even created" [75]. This proactive stance aligns closely with the goals of anticipatory design: not just responding to change but helping shape it.

Future studies　　　　*Foresight*　　　　*Forecasting*

FIGURE 14.1　Layers of future exploration.

Forecasting addresses short-term probabilities; foresight expands into broader possibilities; future studies encompass the full spectrum of uncertainty and opportunity. For anticipatory design, this distinction is critical. While forecasting is valuable for short-term planning, foresight is necessary to design relevant, resilient, and ethically grounded systems over time.

14.3 FORESIGHT TOOLS AND TIME HORIZONS

Anticipating user needs isn't about guessing the future—it's about preparing for it. When we design for today's users, we mostly look at present behavior and needs. When we design anticipatory systems, we are working across time horizons:

- **Past**: Learning from historical patterns to guide our assumptions.
- **Present**: Reading signals in today's environment to spot early changes.
- **Future**: Envisioning not just one future, but multiple possibilities users might encounter.

But how do we map futures that have yet to happen? The Futures Cone is one of the most powerful models for guiding our thinking.

Most organizations plan as if the future is a straight line: project trends, extrapolate, and prepare. In reality, the future is a cone that widens over time, full of diverging possibilities:

- **Potential futures**—*Future Knowledge "might happen"*—Everything that could possibly happen, even wild and implausible scenarios.
- **Plausible futures**—*Current Knowledge "could happen"*—Futures that could happen, grounded in current understanding of how the world works.

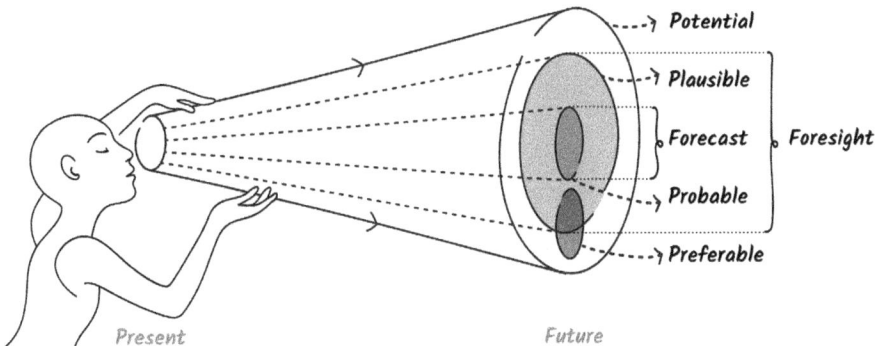

FIGURE 14.2   Future cone diagram of types of alternative futures [76].

- **Probable futures**—*Current Trends "likely to happen"*—Futures likely to happen, usually based on current trends.
- **Preferable futures**—*Value Judgments "want to happen" or "should happen"*—Futures we desire to happen, reflecting our values and aspirations.

The Futures Cone reminds us that designing for the future isn't just about responding to likely outcomes. It's about intentionally shaping preferred ones.

14.3.1 Building Your Foresight Toolkit

Just like we have tools for user research, prototyping, and usability testing, you can also use tools to explore futures. Here are the essential ones for a user experience foresight toolkit [75]:

1. Scenario Planning

Imagine a range of plausible futures—not just the most likely one. Scenario planning helps teams explore how different environmental, technological, or social shifts might unfold, and design systems flexible enough to adapt.

Use scenarios to pressure-test your designs:

- What happens if user trust in AI collapses?
- What if a new regulation changes data privacy overnight?

Scenarios help design teams prepare for surprises, not just predict trends.

2. Horizon Scanning

Identify early signals of change—weak signals. This method systematically scans industries, technologies, demographics, politics, and culture to detect emerging patterns before they become obvious. It's like **future sensemaking**: what quiet signals today could become loud changes tomorrow?

3. Trend Mapping

Connect early signals into meaningful patterns. Trend mapping helps teams see how scattered weak signals, emerging technologies, and shifting user behaviors could converge into larger movements. By mapping trends, designers can move beyond reacting to isolated changes and proactively shape how systems evolve to meet emerging user needs.

4. Weak Signal Analysis

Instead of focusing only on mainstream behaviors, pay attention to early adopters, niche communities, or subtle changes. Weak signals often look small now, but can grow into disruptive shifts. Today's fringe behavior might be tomorrow's mainstream.

5. Visioning
Start with the future you want to create, backcasting:

- What must be true for this future to exist?
- What design actions can we take today to make it real?

Visioning ensures that we don't just passively adapt to change but actively shape it.

6. Why These Tools Matter for Anticipatory Design?
These foresight tools don't replace core UX research methods. They expand them, helping us see user needs not just as they are today but as they might become tomorrow. Without foresight, anticipatory systems risk designing for futures that never materialize or, worse, creating systems that alienate users as reality changes. With foresight, you can build resilient, meaningful, and human-centered systems even as the world shifts around them.

14.4 THE MISSING LAYER OF ANTICIPATION

As designers working with future-facing systems, it's tempting to believe that forecasting and foresight are enough. After all, forecasting helps us project likely outcomes based on the past, while foresight helps us prepare for a range of possible futures. But building truly resilient anticipatory systems requires something more: a present-oriented capability that moves beyond prediction and preparation.

This chapter builds on Roberto Poli's insights, arguing that anticipation should be treated as a third, essential dimension of future studies, distinct from forecasting and foresight [77].

> *Forecasting, Foresight, and Anticipation are considered to be the basic components of Futures Studies. Forecasting deals with data extrapolation; Foresight with the visualization of possible futures; and Anticipation with the translation of their outcomes into action.*
>
> —*Roberto Poli [78]*

In other words, Poli's theory adds a missing layer to future studies: while forecasting and foresight help us *think* about the future, anticipation helps us *act* on it in the present.

However, for Poli, forecasting merely represents a fraction of future studies. Poli and Fasoli emphasize that forecasting and foresight are models—representations of possible outcomes and patterns. Anticipation, however, addresses a different challenge: **How to translate future possibilities into concrete decisions today?** [77]. Drawing inspiration from Poli's theory of anticipation [79], we can visualize how anticipation complements the future thinking process in Figure 14.4.

- **Forecasting** extrapolates trends from historical data.
- **Foresight** explores and visualizes multiple possible futures.
- **Anticipation** transforms this knowledge into present-day action.

Anticipation is the ability to act today based on the recognition of emerging futures—whether or not they can yet be precisely forecasted.

—Roberto Poli [79]

Rather than trying to predict the future with certainty, anticipation acknowledges complexity and uncertainty, and treats them as sources of design opportunity.

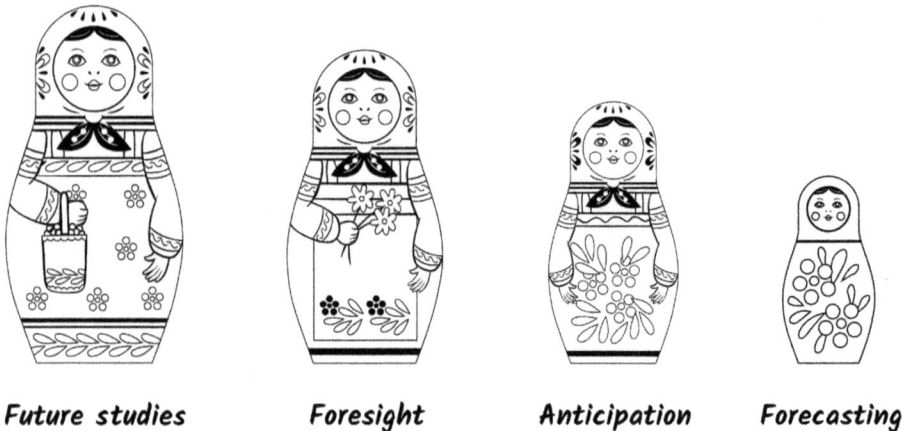

Future studies   **Foresight**   **Anticipation**   **Forecasting**

FIGURE 14.3 Relationship between forecasting, foresight, and future studies.

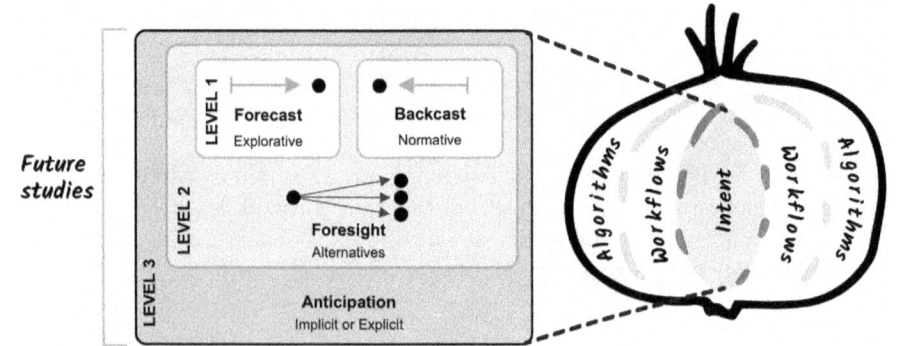

FIGURE 14.4 My understanding of Poli's theory concerning the three levels of future studies [79, 80].

14.4.1 Why Anticipation Matters for Anticipatory Design?

In the context of anticipatory systems, this matters enormously.

- **Forecasting** helps predict what is *likely* to happen.
- **Foresight** helps envision *what could* happen.
- **Anticipation** ensures that today's design choices align with both the possible and the preferable futures.

Without anticipation, anticipatory systems risk becoming rigid—optimized for yesterday's assumptions instead of adapting to the realities of an evolving user landscape. Anticipation invites us to treat the future not as a destination to predict but as a landscape we can influence through every decision we make today. It closes the loop between **insight** and **action**.

Integrating foresight and anticipation into anticipatory design significantly improves our ability to navigate uncertainty, complexity, and what Rittel and Webber called "wicked problems"—multifaceted challenges that defy easy solutions.

Some of the key advantages include the following:

- **Temporal alignment**: Foresight bridges past, present, and future. Anticipation turns this into informed action.
- **Scenario planning**: Designing multiple pathways enables systems to flex as real-world conditions shift.
- **Visioning and backcasting**: Keeping systems anchored to human values, not just technological capabilities.

By weaving foresight and anticipation into design practice, we move from building systems that merely react to trends to creating **future-resilient, human-centered, and ethically grounded systems**. This interdisciplinary approach—combining design strategy, systems thinking, and futures literacy—will be essential if we want anticipatory systems to serve not just users' needs today but also the evolving, complex realities of tomorrow.

14.5 APPLYING FORESIGHT IN DESIGN PRACTICE

So far, we've explored why foresight is essential to anticipatory design and how it expands our capacity to imagine beyond predictable trends. But understanding the theory is only half the work. The real challenge, and opportunity, lies in **translating foresight into everyday design practice**. Let's shift from mindset to methods: how designers can embed foresight into UX, product, and service design to create systems that don't just respond to change but actively shape it.

14.5.1 Designing Beyond the Next Feature: Thinking in Horizons

Traditional design often orbits around the next sprint, the next release, and the following six-month roadmap. Foresight challenges us to think differently, such as:

- What will our users' goals look like one, five, or ten years from now?
- How might societal, technological, or environmental shifts reframe the problems we are trying to solve?
- How can we design systems flexible enough to evolve alongside users, not just deliver today's solutions?

Practical Action: Include "far-future" workshops in your design cycles when setting product vision or defining platform strategy. Use the Futures Cone to map **plausible, probable, and preferable** user futures—and challenge assumptions about what possible requirements your service will need to fulfill.

14.5.2 Envisioning Emerging User Archetypes

Today's personas and journey maps reflect present behaviors and attitudes. But what if those behaviors change? What new user archetypes might emerge as technology, demographics, and cultural norms evolve?

For instance, the rise of AI literacy, climate anxiety, or decentralized economies could fundamentally reshape users' expectations about trust, privacy, or ownership. If we design only for today's personas, we risk becoming irrelevant tomorrow.

Practical Action: Expand user research to include "future persona" exercises.

- Imagine users who grew up with AI as a default.
- Model users navigating hybrid physical–digital worlds.
- Prototype services for users facing environmental instability, new work models, or alternative education systems.

Anticipating these new user archetypes makes products more resilient, inclusive, and meaningful over longer time horizons.

14.5.3 Using Scenario Planning and Backcasting for Roadmaps

Roadmaps typically plot a linear sequence of features and releases. But what if the environment changes? What if new regulations, competitor behaviors, or even environmental catastrophes upend the plan? Scenario planning and backcasting offer powerful anti-dotes to rigid roadmaps:

- **Scenario planning** explores multiple plausible futures, so we are not locked into one fragile path.
- **Backcasting** starts from a preferable future and maps how to work backward from there, ensuring that day-to-day decisions stay aligned with longer-term goals.

Practical action: Pair quarterly milestones with scenario-based contingencies when defining product roadmaps. Ask the following:

- What if a significant technological disruption occurs?
- What if user trust declines in automated services?
- What if sustainability regulations force design pivots?
- What if economic instability (like tariff wars, inflation, or supply chain disruptions) reshapes user behaviors or market dynamics?

By practicing backcasting, teams can stay future-aligned without becoming future-blind.

14.6 FORESIGHT IS NOT A ONE-TIME EXERCISE

Finally, it's important to remember: foresight isn't something you do once at a kickoff meeting. It's a practice, an ongoing process of scanning, questioning, envisioning, and adapting. Incorporating foresight into design practice means regularly asking the following:

- What assumptions are we making about users or the world that might no longer hold true?
- What signals of change are emerging that might reshape our solution space?
- How can we design flexibility, choice, and resilience into what we build?

Foresight-infused design doesn't promise certainty. It offers better **preparedness**, **adaptability**, and **alignment with evolving human and world needs**. Designers who develop these skills won't just create better user experiences—they'll build systems capable of evolving with people, contexts, and the uncertainties of the future.

TL;DR

Forecasting predicts. Foresight prepares. Anticipation acts—Designing truly anticipatory systems demands more than projecting from the past. It requires preparing for multiple futures and translating possibilities into present-day action.

- **Forecasting** helps designers project what is likely based on historical data and trends.
- **Foresight** encourages exploration of what could happen, acknowledging that futures branch, diverge, and surprise.
- **Anticipation** focuses on taking informed actions today, recognizing early signals of change and adapting dynamically.

Without foresight, anticipatory systems become fragile. Systems risk optimizing for yesterday's behaviors, missing edge cases, and failing to adapt to user needs, technologies, and environmental shifts.

Foresight fills a critical gap in traditional design education—Design frameworks like Design Thinking and Double Diamond were built for stable problem spaces. Today's complexity demands that designers develop futures literacy, strategic flexibility, and ethical foresight.

Key foresight methods for designers:

- Futures Cone
- Scenario Planning
- Horizon Scanning
- Trend Mapping
- Weak Signal Analysis
- Visioning and Backcasting

Poli's theory of anticipation adds the missing layer—Forecasting and foresight model possibilities; anticipation transforms them into actionable present-day strategies.

Foresight isn't a one-time exercise—It's a continuous design practice that questions assumptions, scans for change, envisions emerging needs, and adapts systems over time.

TABLE 14.1   Forecasting Versus Foresight Versus Anticipation in Anticipatory Design.

APPROACH	FOCUS	PURPOSE	STRENGTHS
Forecasting	Past and present data trends	Predict likely short-term outcomes	Data-driven; useful for stability
Foresight	Plausible and possible futures	Explore multiple scenarios to prepare for change	Resilient thinking; embraces complexity
Anticipation	Present action informed by emerging futures	Guide immediate design decisions to align with evolving contexts	Real-time adaptability; bridges the future and the now

PART IV

Designing Anticipatory Systems

How Foresight Integrates With Traditional UX Workflows?

15

Building on the previous chapter's foundation in strategic foresight, this chapter brings anticipatory thinking into the everyday practice of UX. If foresight helps us imagine what might come next, this chapter explores how to design for it. It revisits the Double Diamond and shows how future-facing methods can expand each phase. In doing so, this chapter offers a bridge between vision and execution, making it possible to design anticipatory systems that are not only strategically grounded but also operationally actionable.

Earlier in the book, we introduced the Double Diamond framework as a foundational model in design thinking. Due to its simplicity and clarity, it remains one of the most widely adopted frameworks in UX practice. Its four phases—**discover, define, develop, and deliver**—guide teams through divergent and convergent thinking cycles, from exploring the problem space to crafting and deploying solutions.

Traditionally, the Double Diamond centers on current user needs and near-term feasibility. But anticipatory design challenges us to go further—to ask not just "What do users need now?" but also "What will they need next?" This shift demands a broader temporal lens. By integrating foresight methods into each phase of the Double Diamond, we transform the model into a powerful tool for designing systems that remain relevant, resilient, and ethical in an uncertain future.

DOI: 10.1201/9781003642800-19

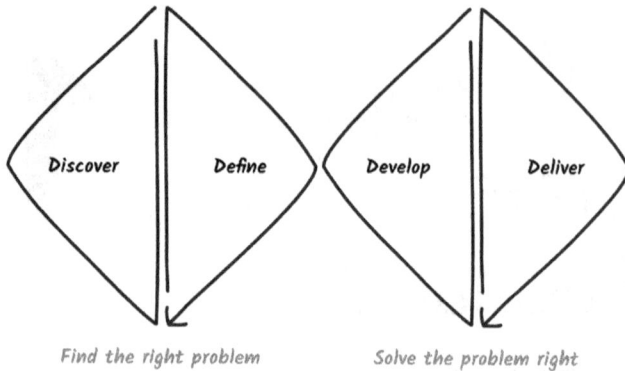

Find the right problem   Solve the problem right

FIGURE 15.1 The double diamond process.

15.1 REFRAMING THE DOUBLE DIAMOND WITH A FORESIGHT LENS

Traditionally, the Double Diamond phases focus on the present and immediate user needs. However, incorporating foresight significantly enhances this process, enabling designers to create user experiences that not only are relevant today but also may thrive in the years to come. By integrating foresight methods into each phase of the double diamond, we move beyond reactive problem-solving and embrace a proactive approach to design. Let's explore how foresight aligns with and extends each phase of the Double Diamond:

15.1.1 Discover—Scanning the Horizon

In the traditional discover phase, teams gather insights about current user behaviors, pain points, and market context. When augmented with foresight, this phase should include foresight tools like **horizon scanning**, **trend analysis**, and **weak signal detection**. These methods help uncover early indicators of change—technological, social, environmental, economic, and political—that may influence future user needs.

Foresight Tools
- Social, Technological, Economic, Environmental, Political (STEEP) analysis
- Trend matrices
- Expert interviews

Key question: What signals are emerging today that may shape the user experience of tomorrow?

For instance, a team designing a city transportation app might not only research current commuting patterns but also consider the potential impact of self-driving cars, the rise of micromobility, or changes in urban planning. This anticipatory approach allows for identifying unmet future needs, leading to more innovative solutions.

15.1.2 Define—Framing Futures

Once the discovery phase is complete, the next step is defining the problem. A traditional approach involves summarizing the user needs and pain points identified in the discovery phase. Incorporating foresight into this phase means defining the problem not only in its current context but also in the context of potential future scenarios.

In this phase, insights are synthesized to define the core problem or opportunity. Foresight methods enhance this by reframing design challenges in light of future uncertainties. **Scenario planning** and **causal layered analysis** help teams explore multiple futures and understand deeper drivers behind surface-level problems.

Foresight Tools
- Eight-step scenario planning
- Futures wheel
- Three horizons framework

Key question: How might this problem manifest in different futures—and what problem are we *really* solving?

For instance, if the transportation app team anticipates a significant shift toward self-driving cars, their problem statement should reflect this, perhaps focusing on how to design an app that seamlessly integrates with autonomous vehicle systems. This foresight-informed definition ensures that the design solution is adaptable and resilient to future changes.

15.1.3 Develop—Designing With Time in Mind

This phase emphasizes ideation and prototyping. With a foresight lens, teams can co-create speculative concepts for different future contexts. **Backcasting** is particularly valuable here, starting from a preferred future and working backward to define the necessary steps to achieve it.

Instead of designing for a single, predictable future, the team develops multiple prototypes, each tailored to a specific scenario.

Foresight Tools
- Backcasting workshops
- Design fictions and speculative artifacts
- Temporal design probes

Key question: What would a desirable future look like, and what must we build today to make it possible?

For the transportation app, this could involve creating different versions of the app—one optimized for use in a world with widespread autonomous vehicles, another designed for a future with increased reliance on micromobility, and perhaps another for a scenario where public transport has undergone significant improvements. This approach allows the team to assess the adaptability and resilience of their design across various potential futures. This iterative process of prototyping and testing helps to refine the design, ensuring that it remains relevant and useful despite future uncertainties. Using speculative design allows for exploring radical future possibilities and encourages the team to think creatively about the app's potential.

15.1.4 Deliver—Prototyping for Uncertainty

The final delivery phase usually focuses on launching the product or service. However, a foresight-informed approach prioritizes a more iterative and adaptable delivery. Instead of viewing the launch as a fixed endpoint, the design team integrates mechanisms for continuous monitoring, feedback, and adaptation. This means designing for ease of update, allowing for adding or modifying features as future trends and user needs evolve. **Stress-testing across scenarios**, **resilience mapping**, and **anticipatory KPIs** ensure that designs hold up in volatile futures.

Foresight Tools
- Wind tunneling
- Scenario-based usability testing
- Leading indicator metrics

Key question: How resilient is our solution across various plausible futures?

For the transportation app, this could include a modular design that allows for effortless integration of new transportation options or features as they emerge in the future. Regular user feedback loops and data analysis become critical in this phase, enabling the team to fine-tune the app based on real-world usage data. This continuous improvement cycle is crucial in a rapidly changing technological landscape.

15.1.5 Why This Matters?

The integration of foresight doesn't simply add another step; it fundamentally alters the nature of the design process. It fosters a more proactive, resilient, and adaptable approach, transforming the designer from a reactive problem-solver to a strategic anticipator of future needs and trends. This shift is especially important in rapidly changing fields like technology, where user behavior and expectations constantly evolve.

By embedding foresight into the Double Diamond, we shift from designing *for the moment* to designing *through time*. This expanded approach prepares us not only to meet current expectations but to shape meaningful, future-ready experiences that evolve alongside users and societies.

15.2 COMMUNICATING FORESIGHT INSIGHTS TO STAKEHOLDERS

Designing anticipatory systems requires not only looking ahead but also helping others see what you see. While foresight methods like trend analysis and scenario planning can illuminate plausible futures, they often fail to create momentum within an organization unless their insights are clearly communicated, emotionally resonant, and actionable. This is where many anticipatory efforts stall: not in the thinking but in the translation.

Earlier in the book, we discussed how designers bridge the gap between technical systems and human needs. In the same way, designers must now act as translators between foresight outputs and stakeholder priorities. Whether you're aligning with executives, collaborating with engineers, or onboarding product teams to speculative work, anticipatory design depends on how well these futures are *framed*, not just envisioned.

Two communication techniques are particularly effective for bridging foresight and strategy: scenario personas and strategic storytelling.

15.2.1 Scenario Personas

While personas are a staple of UX, scenario personas introduce temporal depth. Instead of representing a static archetype, they describe how the same user might evolve across different future contexts. For instance, how might an older adult manage healthcare in a future dominated by wearable diagnostics? What happens if that same future is also shaped by widespread economic inequality?

By layering foresight onto personas, teams can design not just *for* users but *with* an awareness of how their needs may shift under different societal, technological, or environmental conditions. This helps stakeholders see the stakes of today's design choices and encourages more robust, adaptable solutions.

15.2.2 Strategic Storytelling

Data alone rarely drive alignment, but stories do. Translating foresight insights into compelling narratives helps stakeholders internalize potential futures and see where they fit. Instead of abstract trend charts, you might tell a story: a teenager navigating a hyperpersonalized learning system, or a city planner adapting transportation policy based on predictive climate modeling.

Well-crafted foresight storytelling blends emotional resonance with strategic relevance. It connects the dots between today's decisions and tomorrow's realities. The goal isn't prediction—it's readiness. Stories create a shared vision, which is essential for buy-in.

Ultimately, communicating foresight isn't about simplifying complexity—it's about *humanizing* it. Stakeholders don't need to become futurists; they need to see that foresight can clarify uncertainty, reveal opportunity, and build systems that remain resilient over time.

By giving anticipatory insights a shape and a story, we increase the chances they will inform not just what we build but also why we build it.

15.3 FUTURE-ORIENTED USER RESEARCH TECHNIQUES

Throughout this book, we've established that anticipatory design requires more than short-term prediction—it demands a long view, grounded in user intent, behavioral shifts, and the dynamic evolution of context. Yet most traditional user research methods are optimized for present-tense needs: current pain points, observable behaviors, or near-future usability. So, how can we study *what doesn't exist yet*? How do we investigate user experiences that may never happen, but could?

Future-oriented user research is the missing link between speculative foresight and grounded UX strategy. It adapts established qualitative methods—like interviews, ethnography, and surveys—for anticipatory goals. The shift is not in abandoning rigor but in reorienting our lens: from *what is* toward *what might be*. These adaptations enable teams to study imagined contexts, probe future intent, and surface the values and behaviors that could emerge under shifting technological or societal conditions.

This section explores how to evolve familiar tools into future-facing methods that retain their methodological strength, while unlocking new creative and strategic insights.

15.3.1 Ethnographic Probing in Future Contexts

Traditional ethnography focuses on contextual inquiry: observing people in real environments to understand implicit needs and cultural behaviors. In future-oriented ethnography, the "field" becomes speculative. Instead of just shadowing today's workflows, researchers can embed provocations—for example, speculative artifacts, future scenarios, or "what-if" probes—to explore how behaviors might change under alternative conditions.

Example Applications
- Place a mock future interface or AI-enabled artifact into a user's routine and observe not how they use it, but how they adapt to it, resist it, or reconfigure its meaning.
- Conduct day-in-the-life futures, asking participants to narrate or role-play their routine 10 years from now with a new class of tools or societal shifts in place.

This method builds on an existing ethnographic discipline but extends it into a temporal simulation, treating behavior not just as data but as rehearsal.

15.3.2 Future Interviews and Temporal Projective Techniques

In-depth interviews are a core research method for understanding motivations, beliefs, and expectations. But when applied to the future, asking users "What would you want?" or "What will you do?" often leads to shallow, speculative answers. The key is to design **future-tense prompts** that shift perspective and unlock deeper narratives.

Useful Adaptations Include
- **Time travel interviews**: Ask participants to speak from a future point of view—"It's 2035. Your smart health advisor knows your habits better than your partner. What does your day look like?" Framing the future as a lived reality, not a distant prediction, invites more grounded detail.
- **Artifact-input and Wizard of Oz**: Introduce a speculative product or service (a voice-enabled grocery contract, a predictive hiring app) and ask users to respond to it emotionally and ethically, not just functionally. What would they trust? What would they fear? This method becomes even more powerful when paired with **Wizard of Oz** technique, where users interact with what they believe is an autonomous system, but which is actually being operated manually behind the scenes. By simulating the anticipatory behavior without full technical implementation, researchers can observe real-time user reactions, hesitation points, and misalignments between system behavior and user expectations.
- **Success/failure backcasting**: Ask participants to imagine a future system that succeeded or failed, and then walk backward through what led to that outcome. This can help reveal unspoken values, triggers of distrust, and design risks early.

These interviews aim less at "getting it right" and more at **mapping possibility spaces**, highlighting latent hopes and unresolved tensions about the future.

15.3.3 Foresight-Aligned Survey Design

Quantitative methods like surveys can also adapt to a foresight lens by using **scenario-driven questions** and **alternative future framing**.

For Example
- Rather than asking "How likely are you to use X?", ask "If data-sharing norms changed in five years, how comfortable would you be with."
- Use **scenario-based choice models** to test behavioral shifts across multiple futures: for example, "In a world where health data is fully public, which platform would you trust?"

Designing surveys for **temporal sensitivity** helps validate how attitudes and preferences may shift across plausible worlds. You can also incorporate **Delphi-inspired iterations**, asking respondents to reconsider their answers after being shown what others predicted, helping simulate the push and pull of societal change.

15.3.4 Future Personas and Temporal Mental Models

Personas are a powerful way to synthesize user research. In anticipatory design, **scenario personas** can be used not just to represent current users, but *evolving* users under different social, environmental, or technological conditions.

Additionally, map **mental models** not as fixed beliefs but as **cognitive trajectories**:

- What might a user believe *before* they adopt an AI-driven assistant?
- How will their mental model shift *after* three months of usage in a highly adaptive system?

Future mental models only are speculative, but they help design for **cognitive onboarding**, trust calibration, and long-term alignment, especially for systems that require a change in user expectations over time.

In a world where designers are expected to operate beyond the moment, these research techniques provide both imagination and rigor. They ensure that we don't just project technology forward but also bring **people** with us into the future we're designing for.

TL;DR

Reframing the double diamond for foresight—Foresight transforms the Double Diamond from a discovery process into a strategy of anticipation:

- **Discover** becomes horizon scanning—where research expands to include emerging signals, trends, and long-term shifts.
- **Define** becomes future framing—where problems are situated not just in the now, but across possible, probable, and preferred futures.

- **Develop** becomes temporal prototyping—using backcasting and speculative methods to build for what might be.
- **Deliver** becomes resilience testing—ensuring that designs can adapt, evolve, and endure across uncertainty.

Without foresight, we risk designing experiences optimized only for the short term, solutions that age quickly, misalign with emerging needs, or fail under pressure.

Foresight adds temporal depth, strategic resilience, and future readiness to our most familiar frameworks. It turns the Double Diamond into a prism, bending linear processes into multidimensional exploration.

Communicating foresight insights to stakeholders—Foresight loses power if it stays locked in strategy decks. To shape anticipatory systems, designers must also shape how futures are understood and shared.

- **Scenario personas** extend traditional personas by embedding users in alternative futures.
- They help teams consider how user needs, constraints, and behaviors may evolve under different societal, technological, or environmental conditions.
- **Strategic storytelling** translates foresight insights into emotionally compelling narratives.
- By anchoring future scenarios in lived experience, storytelling helps stakeholders visualize the consequences and opportunities of today's decisions.

Clear communication makes foresight usable. It aligns teams, inspires buy-in, and ensures that anticipatory design becomes a shared endeavor, not just a design vision but also a strategic act.

Designing Anticipatory Experiences

16

As AI systems become more anticipatory and autonomous, the role of user experience design is evolving from reactive usability to proactive coordination. This chapter introduces a comprehensive framework to assist both designers and non-designers in creating intelligent systems that responsibly predict, adapt, and guide user behavior. Drawing on behavioral science, foresight, and anticipatory design principles, this chapter presents a structured, iterative model that equips teams to build experiences that balance automation with user agency, predict needs while respecting autonomy, and create value without overwhelming the user. The framework comprises three core design anchors: user intent, behavioral science, and temporal foresight. Each phase is structured using UX Chunks, ensuring that the design process is actionable, comprehensible, and cognitively manageable. Actionable methods, designer mindsets, and example-driven guidance accompany each phase.

The shift from reactive tools to proactive agents is redefining the UX discipline. No longer just about usability, anticipatory design now demands a deeper coordination between system behavior and human intent. Anticipatory systems aim to reduce friction, predict needs, and guide users through decisions before they even arise. Done well, these systems enhance agency and simplify complexity. Done poorly, they risk manipulation, opacity, or loss of trust.

The framework presented in this chapter integrates methodologies from strategic foresight and future studies, presented in Chapter 10. These techniques allow designers to envision diverse futures, navigate uncertainties, and align short-term predictions with long-term goals. Beyond facilitating immediate decision-making, anticipation serves as a bridge between the present and the future. This chapter synthesizes qualitative and quantitative methodologies to explore how anticipatory systems can harmonize probabilistic predictions (forecasting) with goal-oriented strategies (backcasting). These approaches equip designers and non-designers to tackle complex challenges related to user intent, workflows, and algorithmic optimization.

Each phase is built on three design anchors—**user intent, behavioral science**, and **temporal foresight**—and structured using **UX Chunks** to ensure cognitive manageability and real-world applicability.

16.1 WHY NOW?

As AI becomes more autonomous, the boundary between system behavior and user agency begins to blur. Anticipatory design matters now because today's AI doesn't just respond to user input—it predicts it. That power changes the nature of interaction. UX professionals are no longer just optimizing flows or reducing clicks; they're shaping the decisions users never explicitly make. This shift places new responsibilities on UX designers and teams working with AI-driven systems:

- How do we align predictive systems with human intent rather than just behavioral patterns?
- How can we design for proactive engagement without undermining autonomy?
- How do we embed transparency, accountability, and adaptability into evolving systems?

This framework offers a design-centered response. It distills a decade of cross-disciplinary research, behavioral science, and field experimentation into a practical methodology for crafting anticipatory systems that are helpful, transparent, and human-centered.

Whether you're designing a health app that nudges patients toward healthier routines or an autonomous platform that adapts to learner progress, this framework offers the tools to balance intelligence with empathy, automation with agency.

16.1.1 Structuring the Framework With UX Chunks

To translate cross-disciplinary ideas, I adopted *UX Chunks* approach to help translate complex anticipatory design tasks into structured, manageable actions for both stakeholders and design teams. *UX Chunks* is a concept adapted from George A. Miller's research on working memory [81]. His foundational insight—that most people can manage about seven pieces of information simultaneously—has shaped decades of interface design. I adopt this principle to systems design.

Organizing anticipatory systems into digestible "chunks"—manageable phases, defined roles, and targeted actions—makes complexity easier to grasp. These chunks stem from the following:

- **Cognitive load theory**: Reduce mental fatigue by sequencing complexity.
- **Interaction design**: Break tasks into progressive, learnable steps.

- **Information architecture**: Structure content hierarchically to guide attention and action.

The proposed framework translates these insights into a concrete, designer-friendly methodology for crafting anticipatory systems. This commitment to innovation, ethical responsibility, and user empowerment ensures that anticipatory systems elevate, rather than undermine, the human experience. The proposed framework is designed to navigate complexity while maintaining coherence, adaptability, and a user-first perspective, laying a foundation for both theoretical insight and practical application.

16.2 FOUNDATIONS OF THE FRAMEWORK: THREE DESIGN ANCHORS

This framework builds on a comprehensive scope review conducted during my doctoral research, which examined over 50 anticipatory AI experiences across industries. Patterns that emerged from that review directly shaped the principles outlined in this book.

To build intelligent systems that users can trust and adopt, the design process is grouped into three interdependent anchors:

1. Behavioral Science as a Guide
Behavioral frameworks such as the TTM and Nudge theory (Chapter 9) help us shape interventions that resonate with users' motivations and cognitive states. These models allow us to design support systems that nudge users toward healthier or more efficient behaviors while respecting their autonomy.

2. User Intent as a Central Focus
Understanding user intent is essential in highly automated systems. As explored in Chapters 12 and 13, mental models shape how users interpret system actions. Systems must learn to align with both stated and latent intent, using research techniques like jobs-to-be-done (JTBD), automation mapping, and empathy interviews.

3. Temporal Integration
Forecasting, backcasting, and foresight tools (Chapter 14) allow designers to build systems that are responsive in the short term but aligned with long-term human and societal goals. This temporal thinking helps anticipatory systems evolve with the people they serve.

16.3 THE ANTICIPATORY UX FRAMEWORK: A THREE-PHASE PROCESS

Anticipatory design requires a multidimensional approach that incorporates temporal methods (forecasting, backcasting, and foresight), human-centered principles, and ethical considerations. Consequently, you'll explore a modular framework broken into three actionable phases:

- **Anticipate**—*Anticipating Directions of Change*—Investigate trends, behaviors, and signals to understand what's coming next.
- **Imagine**—*Imagining Alternative Scenarios*—Envision alternative futures and prototype ethical, explainable systems.
- **Shape**—*Shaping Choices*—Build and refine anticipatory experiences that balance automation with control.

Each phase is paired with designer mindsets, real-world methods, and sample activities, offering a toolkit for responsible innovation. By the end of this chapter, you'll have a roadmap for building anticipatory systems that don't just work but work *with* people. Systems that are not only predictive but participatory. Not only intelligent but intelligible.

Each module contains specific yet adaptable procedures, breaking down anticipatory design processes into a concise set of chunks. The framework aims to guide professionals along a structured and less ambiguous path. By organizing the workflow into meaningful, streamlined phases, professionals can more easily address, understand, and recall each step, enhancing both the efficiency and clarity of their work in creating anticipatory systems.

Each phase is designed to increase engagement and creative relevance:

- **Anticipate = investigative mindset**: Observe patterns, question assumptions.
- **Imagine = speculative mindset**: Envision alternatives, embrace ambiguity.
- **Shape = iterative mindset**: Test, refine, and remain open to feedback.

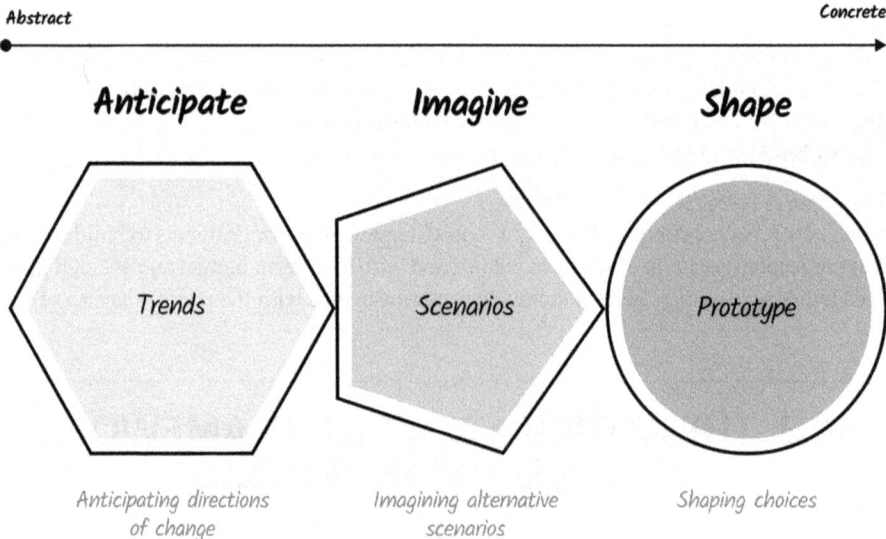

FIGURE 16.1   The anticipatory design framework proposed in the book.

16.4 PHASE 1: ANTICIPATE

Anticipatory design begins with awareness of user needs, behavioral shifts, and the larger forces shaping future experiences. This phase involves making sense of change—collecting signals, understanding intent, and mapping what matters most.

Anticipate is structured into three UX Chunks—**exploration, synthesis**, and **vision**—each supporting designers in uncovering and organizing insights without overwhelming cognitive load. These sub-phases combine qualitative research, trend analysis, and foresight methodologies to help design teams anticipate—not just react to—evolving user expectations.

The objective is to identify and align with emerging behaviors, motivations, and societal trends to lay the groundwork for adaptive, responsible AI systems. This phase ensures that anticipatory experiences begin with the right questions, not just the right data.

At its core, designers begin **by identifying user intent**—understanding goals, motivations, and pain points. Frameworks like JTBD help uncover the underlying reasons behind user actions and behaviors, ensuring that designs address both immediate and aspirational needs. For instance, a ride-sharing service might analyze demographic shifts, transportation infrastructure changes, and evolving environmental concerns to anticipate future demand and optimize its offerings.

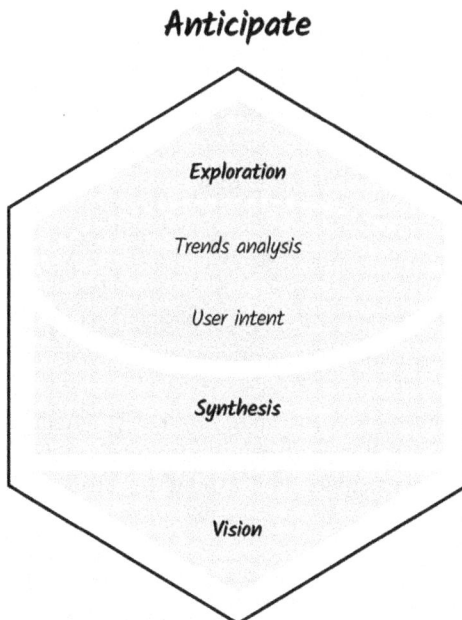

FIGURE 16.2   Overview of phase 1, anticipate.

PA serves as a cornerstone for this phase. Historical user data, combined with external factors like market trends or economic conditions, enable the identification of patterns and predictions about future behaviors. For example, anticipating a surge in demand during a specific event or adjusting pricing models based on predicted changes in fuel costs are proactive measures that stem from robust data analysis. To achieve this, a diverse array of data points must be incorporated, minimizing bias and enhancing accuracy.

Purpose
Identify major trends and forces shaping the future landscape of user needs, motivations, and technological possibilities. By synthesizing qualitative insights, predictive models, and foresight methodologies, this phase ensures that anticipatory systems are grounded in user intent while remaining agile in the face of change. It creates a foundation for designing adaptive systems that effectively anticipate and respond to evolving user expectations.

Design Mindset: Investigative
AI-enhanced research opens new possibilities for anticipatory design. Today's designers can leverage AI persona builders to accelerate the synthesis of user intent, tools that process vast datasets to generate dynamic, evolving user models. These models can simulate future behaviors and surface latent motivations, and align with foresight-informed scenarios. This still requires a critical, investigative mindset to work well: observe deeply, question assumptions, and stay open to weak signals and surprising user narratives.

16.4.1 Exploration

Focus: Run in-depth research to uncover emerging trends, user intent, and behavioral motivations.

Components
A. Trend Analysis
> **Objective**: Identify external forces (technological, social, economic) influencing user needs.
> **Methods**: Use tools like STEEP analysis, horizon scanning, and expert panels to detect emerging trends and weak signals.
>
> • Horizontal scanning is a foresight method for identifying early signals of change. It's a systematic, proactive approach that empowers organizations to not only anticipate and respond to future trends but to actively shape them [82].

B. User Intent
> **Objective**: Deepen understanding of users' mental models, pain points, and aspirations.

STEEP analysis *template*

S	Social
T	Technological
E	Economic
E	Environmental
P	Political

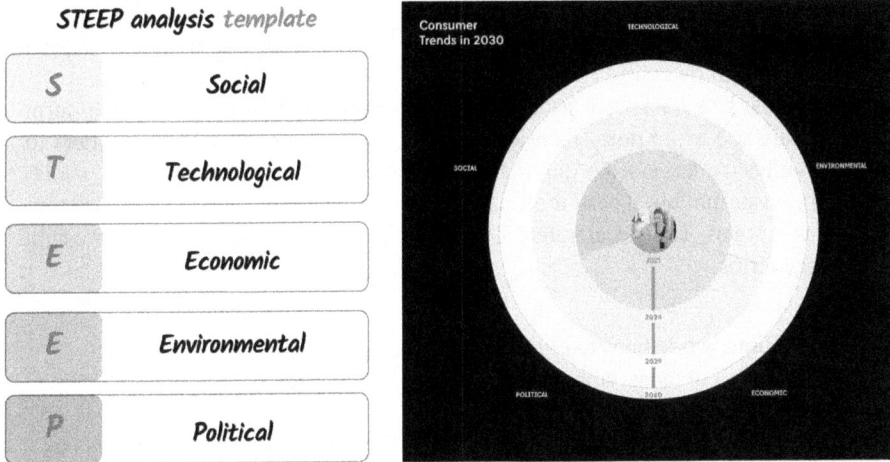

FIGURE 16.3   On the left, a STEEP analysis template. On the right, a foresight radar template for horizon scanning [82].

Methods
- **Empathy mapping**: Visualize user emotions, behaviors, and goals.
- **JTBD**: Clarify underlying user needs and broader contexts.
- **Automation versus augmentation mapping**: Define the spectrum of user autonomy and collaboration with AI.

Activities
- Organize trend workshops with interdisciplinary teams to explore societal, technological, and environmental shifts.
- Conduct user research using interviews, surveys, and observational studies.
- Develop empathy maps and autonomy maps to capture diverse user perspectives.

16.4.2 Synthesis

Focus: Combine insights from research into actionable models and hypotheses.

Components
A. Insight Mapping:
   Objective: Identify patterns and interconnections between trends and user needs.
   Methods: Use affinity diagrams or mind mapping to group research findings.

B. Personas:

> **Objective**: Create detailed personas and context-based scenarios that reflect real-world diversity.
>
> **Methods**: Use data from user research to build personas capturing long-term goals and aspirations. Consider leveraging AI-powered persona builders to synthesize large-scale qualitative and behavioral data into dynamic, evolving user models. These tools can help reveal latent patterns, simulate future user states, and accelerate the creation of personas aligned with foresight scenarios.

Activities

- Facilitate workshops to cluster qualitative and quantitative insights.
- Develop personas emphasizing users' needs, pain points, and interaction goals.
- Align personas and scenarios with identified trends and emerging contexts.

16.4.3 Vision

> **Focus**: Articulate a clear, ethical, actionable vision for future user experiences.

Components

A. Foresight Visioning

> **Objective**: Define a forward-looking vision based on synthesized insights.
>
> **Methods**:
> - **Backcasting**: Map pathways from desired future states to present actions according to user intent.
> - **Delphi method**: Gather consensus among stakeholders on critical design directions.

B. Ethical Considerations

> **Objective**: Ensure alignment between system goals and ethical principles like transparency, inclusivity, and trust.
>
> **Methods**:
> - **Stakeholder alignment**: Collaborate with ethicists, designers, and engineers to define shared principles.
> - **Ethical impact assessment**: Anticipate potential unintended consequences of proposed solutions.
>
> **Activities**:
> - Develop vision boards or roadmaps that represent long-term goals and strategies.
> - Conduct Delphi workshops to align team perspectives on ethical and user-centered priorities.
> - Integrate ethical guidelines into design principles for anticipatory systems.

16.5 PHASE 2: IMAGINE

While the first phase focuses on uncovering current patterns and plausible shifts, the second phase—**Imagine**—centers on deliberately exploring alternative futures. It's a transition from insight to foresight, where design moves beyond projecting trends and starts shaping possibilities.

Imagine is structured into three UX Chunks—**scenario planning**, **alignment**, and **workflows**—each helping teams anticipate diverse user needs, build resilience into design, and architect interactions that can adapt across a range of potential futures.

The objective is to explore plausible and preferable futures to inform adaptive, ethical design decisions. Rather than chasing a single prediction, this phase helps teams prepare for uncertainty by imagining systems that are flexible, human-centered, and grounded in user agency.

Instead of predicting a single "correct" future, this phase investigates multiple plausible and preferable futures to develop systems that can thrive across various contexts. This can require combining robust methodologies like STEEP analysis with innovative techniques like backcasting to craft systems that not only adapt to change but also shape desirable outcomes.

For example, if the desired future is a world with significantly reduced carbon emissions, a transportation app could backcast to identify the necessary features and

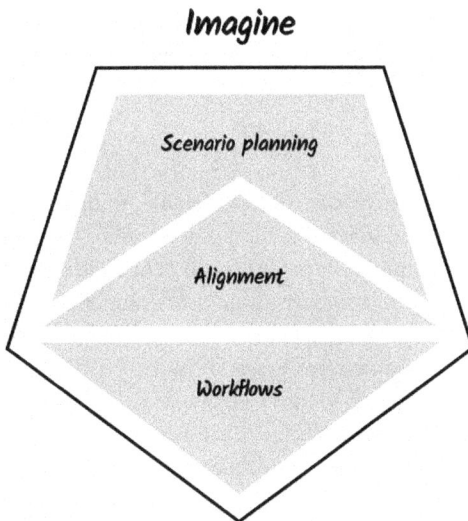

FIGURE 16.4   Overview of the second phase, imagine.

functionalities that would encourage users to choose sustainable transportation options. The diversity of scenarios helps ensure that design decisions are not unduly influenced by a single, potentially inaccurate, prediction of the future. Exploring these alternative scenarios helps prepare the designers to address a wider range of possible future needs.

This phase ensures that anticipatory systems are not only adaptable but also proactive in addressing evolving user needs and societal challenges.

Purpose

Explore various plausible and desirable futures to inform adaptive and impactful design decisions. Rather than predicting a singular future, this phase focuses on creating systems that enhance user control by providing customization options and transparent communication of system decisions, embedding ethical safeguards to ensure privacy, fairness, and accountability, and promoting positive behavior change by integrating nudges and feedback mechanisms that guide users toward their goals.

Design Mindset: Speculative

This phase demands a bold and imaginative mindset. Rather than chasing certainty, designers must embrace ambiguity and use creative constraints to explore systems that support—not script—the future. By designing with flexibility and foresight, teams can prototype interactions that remain meaningful across multiple futures, even when conditions change.

16.5.1 Scenario Planning

Focus: Develop a set of plausible, diverse future scenarios based on external and internal drivers. Assess potential impacts on user needs and how systems could respond to those shifts.

Components

A. Scenario Development

   Objective: Design a set of plausible, probable, and preferable scenarios that account for potential system failures, disruptions, and recovery strategies. These scenarios should reflect real-world challenges and ensure that the system can recover without undermining user autonomy.

   Methods:

   - Combine previously uncovered trends from STEEP and behavioral insights from horizon scanning to identify key factors that could lead to system challenges or failures in user context-based scenarios.
   - **Failure scenarios**: Generate multiple scenarios that include both expected and unexpected failures, designing responses that allow systems to recover without user frustration or loss of trust.
   - **Design for failure**: Explore how different system failures or errors might impact user behaviors and goals, ensuring that users can still control or easily navigate recovery mechanisms.

B. Scenario Analysis

Objective: Analyze how the proposed scenarios impact the long-term user needs and motivations stages.

Methods:

- **Scenario impact assessment**: Assess each scenario's impact on user motivations, behaviors, and interactions.
- **Long-term user needs**: Map long-term user motivations and behaviors in each scenario and ideate nudges that address these needs.

C. Failure Recovery Design

Objective: Plan strategies for system failure recovery, ensuring that the system maintains user trust and autonomy during failures. This includes anticipating how the system should recover from errors without overwhelming users.

Methods:

- **User recovery protocols**: Design clear recovery pathways that users can easily follow, maintaining a sense of control and minimizing cognitive load during failure events.
- **Error versus failure communication**: Define how the system will differentiate between system errors and system failures, and design transparent communication strategies to distinguish between the two to avoid users' confusion and frustration, and define types of explanations.

Activities

- Conduct workshops to generate and analyze a range of plausible, probable, and preferable scenarios.
- Ideate how each scenario would influence the design, focusing on the system's adaptability to each context.
- Develop a scenario matrix to visually compare how scenarios differ and interact.
- Analyze failure points within different scenarios, designing pathways for user support and system recovery.

16.5.2 Alignment

Focus: Start from an ideal future state and backcast to identify the steps required to achieve those outcomes. Ensure that the system design anticipates user needs and recovery processes, focusing on maintaining user autonomy and providing transparent pathways during failures.

Components

A. Backcasting From Desired Future

Objective: Articulate long-term user-centered goals to shape design directions.

Methods:

- **Backcasting workshops**: Co-create future goals with stakeholders (users, designers, technologists). Focus on articulating clear, attainable outcomes.

- **Desired future states**: Define the "ideal" state for users in multiple scenarios and understand their long-term motivations.

B. Use Visioning Planning

Objective: Develop a vision of the future that emphasizes user empowerment, autonomy, and resilience in the face of failure.

Methods

- **Co-creation**: Use collaborative brainstorming sessions with stakeholders to visualize future systems that allow users to navigate potential failures with clarity and control.
- **Future success models**: Define what a successful user experience looks like, ensuring that the user's autonomy, and trust are central to the processes.
  - Emphasize ethical considerations and user empowerment along the journey to respect user autonomy, privacy, and inclusivity, avoiding manipulative tactics.

C. Mapping Milestones

Objective: Reverse-engineer the user journey from the desired future state to identify steps toward achieving those goals.

Methods:

- **User journey mapping**: From the desired future state, map the steps backward to present-day interactions.
- **Backcasting milestones**: Identify critical milestones along the journey and how they relate to system failure recovery, ensuring that users are not overwhelmed at any stage.

Activities

- Conduct workshops with cross-disciplinary teams to co-create and validate the vision for the future system, focusing on user autonomy and failure recovery.
- Map out clear steps to reach desired outcomes, emphasizing transparency and control during failure events.
- Develop roadmaps that include failure recovery touchpoints and ethical guidelines.

16.5.3 Workflows

Focus: Design the user experience by focusing on feedback loops, autonomy protocols, and system explanations. This ensures that the system fosters trust, clarity, and positive engagement while adapting to diverse futures.

Components

A. Feedback Loops

Objective: Ideate adaptive systems that respond to user input over time, building trust through continuous feedback.

Methods:
- **Ideate feedback systems**: Map and design feedback systems that allow the system to evolve based on user inputs and changing contexts.
- **Continuous improvement**: Develop a feedback system that evolves over time based on user input, helping to mitigate future failures and improving system reliability. Based on feedback from users and other stakeholders.

B. Autonomy Protocols
   Objective: Trust calibration. Establish an adaptive design that supports user control without overwhelming them.
   Methods:
- **Autonomy mapping**: Identify decision points and user autonomy levels, ensuring that users maintain control without cognitive overload.
- **User-centric protocols**: Design supporting features (e.g., opt-out choices, transparency, user-driven data management).

C. Reward Functions and Explainability
   Objective: Foster engagement and trust by defining reward functions that incentivize desired behaviors and by designing transparent system explanations.
   Methods:
- **Explainability framework**: Define and implement clear explanation mechanisms across all system touchpoints. Explain the rationale behind system decisions and provide levels of detail based on user needs.
- **Behavioral incentives**: Create reward structures that encourage positive engagement while maintaining transparency and ethical considerations.

Activities
- Ideate adaptive feedback systems that notify users of failure events and guide them through recovery.
- Design user workflows that integrate failure recovery and user autonomy, ensuring that users can act clearly and confidently.
- Run a workshop with data scientists and data engineers to validate how different reward functions and explainability affect user trust and satisfaction.

16.6 PHASE 3: SHAPE

Building on the insights and foresight gathered in the first two phases, the third phase—Shape—focuses on transforming possibilities into tangible, testable systems. It's where concepts become reality, where anticipatory experiences are crafted, challenged, and iterated in context.

Shape is structured into three UX Chunks—**behavioral alignment, prototyping,** and **evaluation**—each helping teams bring adaptive, trustworthy systems into users' hands. This phase emphasizes iterative design, ethical implementation, and feedback-based continuous calibration.

The objective is to leverage scenario insights to guide iterative design decisions. The goal is to build systems that are not only intelligent and predictive but also adaptive, user-centered, and aligned with long-term values.

At its core, this phase involves testing, refinement, and ethical vigilance. It guides the iterative design process, emphasizing agile development, user feedback integration, and ethical considerations to refine the system's design based on scenario insights. This phase ensures that systems are adaptive, resilient, and relevant in real-world applications, allowing for continuous evolution based on real-time user data and feedback.

Systems are evaluated against imagined scenarios to ensure their resilience, adaptability, and relevance in real-world applications. For example, disaster relief systems might undergo simulations to assess performance under various crises. Ethical considerations remain paramount to prevent bias, foster equity, and build trust through transparency and accountability. Behavioral science principles, such as the Nudge theory and the TTM, guide the design of interventions that resonate with users, fostering engagement and long-term adoption.

This phase considers technological advancements and focuses on integrating HCD principles to ensure that the final product is not only functional but also enjoyable and intuitive to use.

Purpose
Leverage scenario insights to inform iterative design decisions, ensuring that systems are adaptive, ethical, and aligned with long-term user goals. This phase focuses on

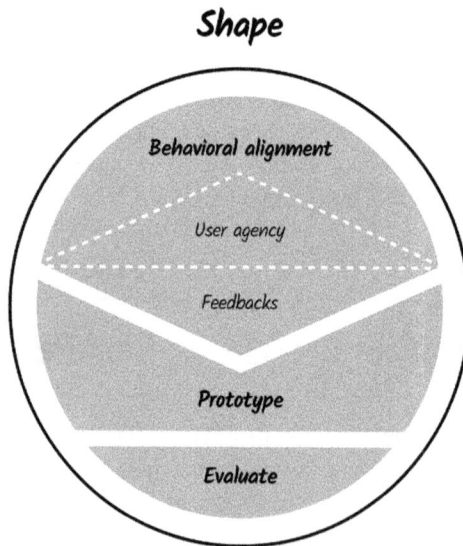

FIGURE 16.5   Overview of the third phase, shape.

enhancing user control, embedding ethical safeguards, promoting positive behavior change, and ensuring that systems are both transparent and responsive. Continuous refinement and ethical vigilance are core to the design process.

Design Mindset: Iterative
Designing anticipatory systems is not a one-shot endeavor—it's an ongoing, responsive practice. This phase calls for an iterative mindset: prototype quickly, test thoughtfully, and design with humility. It's about staying open to adaptation, learning from user interaction, and continuously refining the system in response to real-world conditions.

16.6.1 Behavioral Alignment

Focus: Design for alignment between the system's behavior and user goals, emphasizing trust, autonomy, and adaptability in the face of changing contexts.

Components
A. User Control
   Objective: Prototype interaction control protocols that offer varying levels of autonomy, ensuring that users can modify system behaviors as needed without losing trust.
   Methods:
   • Design interaction protocols for customization of system decisions and automated recommendations.
   • Establish clear communication pathways about the user's control options and transparency around system actions.

B. Designing Feedback
   Objective: Create robust feedback loops to ensure that users understand the system's state and can act accordingly.
   Methods:
   • Prototype system responses that deliver meaningful feedback on user actions.
   • Design recovery mechanisms that are easily accessible, ensuring that users retain control during errors and failures.

Activities
   • Prototype systems with varying levels of user control, from simple adjustments to more advanced customizations.
   • Design adaptive feedback loops, focusing on error recovery and communication of system failures.
   • Test user control protocols to ensure that users can easily modify the system based on real-time contexts.

16.6.2 Prototyping

Focus: Build dynamic, personalized prototypes that foster trust and empowerment through clear, responsive interactions.

Components
A. Personalization
Objective: Develop prototypes that emphasize dynamic personalization and provide users with clear customization settings.
Methods:
- Design onboarding experiences that focus on user preferences and customization options.
- Create adaptive loading states, updates, and system responses that reflect the user's motivations and goals.

B. Building Trust
Objective: Develop prototypes of feedback systems and test them in different scenarios to ensure that users feel confident and in control of the system.
Methods: Implement features that reinforce trust, such as transparency in data usage, explanation of system behaviors, and accessible failure recovery.

Activities
- Develop and test prototypes with clear trust-building elements like user settings, preference management, and error handling.
- Design onboarding interactions that set the tone for transparency and user agency throughout the service lifecycle.
- Test prototypes for user engagement and confidence in system behaviors, ensuring ease of recovery during failures.

16.6.3 Interactive Evaluation

Focus: Use real-time feedback to evaluate and iterate system performance, focusing on evolving user needs, transparency, and explainability.

Components
A. Continuous Testing
Objective: Engage in usability testing and gather real-time feedback to iterate and refine the system based on diverse user contexts.
Methods:
- Run A/B tests, surveys, and usability studies to understand how users interact with the system and where improvements are needed.
- Continuously monitor system performance, addressing issues that impact user trust and satisfaction.

B. Explainability

    Objective: To promote trust and transparency, audit whether the proposed system explanations are clear and context-sensitive to encourage trust and transparency.

    Methods:

- Test explainability features, ensuring that the system clearly communicates the rationale behind automated decisions and offers comprehensible explanations, empowering users to understand and trust the system.
- Focus on providing layered explanations based on user preferences and cognitive load.

Activities

- Conduct usability testing and A/B testing to iterate and refine user interfaces, workflows, and system behaviors.
- Implement and test system explanation features that clearly communicate the rationale behind automated decisions.
- Continuously evaluate the system's ability to adapt to emerging user needs and goals, ensuring that it stays relevant and effective over time.
- Regularly evaluate system performance against KPIs (accuracy, engagement, user satisfaction) and iterate based on these insights.

TL;DR

This chapter introduces a three-phase framework for designing anticipatory experiences. Organized into three phases—anticipate, imagine, and shape—the model integrates behavioral science, foresight, and UX design principles.

Anticipate helps teams identify signals of change, synthesize trends with user motivations, and develop an informed vision. *Imagine* uses scenario planning and backcasting to explore alternative futures and align system workflows with user goals. *Shape* focuses on prototyping, behavioral alignment, and iterative evaluation to build adaptive, trustworthy systems that evolve alongside users. Each phase is structured using UX Chunks to reduce cognitive load and support practical, ethical application.

Design Heuristics for Anticipatory Systems

Turning Principles into Practice

17

As designers begin to operationalize anticipatory design, theory must translate into practice. This chapter introduces a set of heuristics adapted from traditional usability principles to address the unique challenges posed by predictive, data-driven systems. It equips teams with structured guidance to build AI experiences that are transparent, user-aligned, and ethically responsible.

Anticipatory design holds tremendous promise: it aims to reduce cognitive load, predict user needs, and streamline interactions by taking proactive action on the user's behalf. But with that promise comes heightened complexity, risk, and responsibility. As this book shows, designing effective anticipatory systems requires more than smart predictions or elegant automation. It demands clarity, control, and ethical foresight.

While frameworks and theoretical models guide understanding, designers also need **practical tools** to evaluate and iterate on anticipatory experiences. This chapter introduces a set of **heuristics**: structured, experience-based principles that help designers assess and improve AI-powered systems in real-world contexts. These heuristics serve as a bridge between conceptual models and actionable design decisions, empowering teams to make anticipatory systems more trustworthy, usable, and aligned with human values.

With ethical principles, mental models, and anticipatory frameworks now defined, the question remains: *how do we evaluate anticipatory systems in practice?* This chapter offers a pragmatic answer.

DOI: 10.1201/9781003642800-21

17.1 WHY ADAPT HEURISTICS FOR ANTICIPATORY SYSTEMS?

Jakob Nielsen and Rolf Molich's foundational usability heuristics have guided generations of designers in creating intuitive, human-centered interfaces. Their principles—
such as visibility of system status, user control and freedom, and error prevention—
remain timeless. But anticipatory systems introduce new dynamics:

- The user may not initiate actions.
- The system may act autonomously.
- The rationale behind predictions may be opaque.

In this context, traditional heuristics fall short. We need adapted criteria that address:

- Data privacy and transparency
- Overreliance and automation surprise
- Prediction accuracy and feedback loops
- Ethical trade-offs and value alignment

This chapter proposes a set of heuristics tailored to these challenges. It draws on established usability principles while extending them to meet the nuanced demands of anticipatory design.

17.2 ALIGNING HEURISTICS WITH ANTICIPATORY DESIGN CHALLENGES

Table 17.1 maps each of Nielsen's ten usability heuristics to new heuristics specific to
anticipatory systems [83]. Each mapping is paired with a design strategy to overcome
core challenges such as user overreliance, opaque predictions, and limited user control.

TABLE 17.1   Aligning Nielsen's Heuristics With Anticipatory Design Challenges.

NIELSEN'S HEURISTIC	PROPOSED HEURISTICS	DESCRIPTION	STRATEGY
Visibility of system status	Keep users informed	Ensure that users are continuously aware of the anticipatory system's current status and operations.	**Data privacy and security**: Educate users on data privacy policies and practices. **Accuracy and reliability**: Implement feedback loops to improve system performance.

TABLE 17.1   (Continued)

NIELSEN'S HEURISTIC	PROPOSED HEURISTICS	DESCRIPTION	STRATEGY
Match with user expectations	Align with mental models	Anticipatory systems should use language, concepts, and predictions that align with users' expectations and mental models.	**User reliance and overconfidence**: Through design, educate users on the importance of critical thinking.
User control and freedom	Allow user intervention	Provide users with control over anticipatory interactions, allowing them to adjust or override recommendations easily.	**Limited customization and control**: Allow users to adjust recommendation settings and provide feedback.
Consistency and standards	Maintain design coherence	Maintain consistency in design elements and interaction patterns across anticipatory features.	**Accuracy and reliability**: Continuously update and validate models to reflect current data.
Error prevention	Minimize anticipatory errors	Anticipatory systems should include mechanisms to prevent erroneous predictions and actions.	**Accuracy and reliability**: Use high-quality, diverse datasets for training algorithms.
Recognition rather than recall	Reduce cognitive load	Present information and recommendations in context through anticipatory interfaces, minimizing the need for users to remember or anticipate system behavior.	**User engagement and acceptance**: Provide clear, accessible information about how systems work and their benefits.
Flexibility and efficiency of use	Cater to user proficiency	Anticipatory systems should be flexible to accommodate both novice and expert users efficiently.	**Balancing automation and human interaction**: Integrate human oversight in automated processes when critical.
Aesthetic and minimalist design	Present information cleanly	Anticipatory interfaces should be designed clean and uncluttered, focusing on essential information.	**User education and communication**: Develop comprehensive user guides and educational resources.
Error recovery	Assist users in error resolution	Help users understand and recover from errors encountered in anticipatory interactions.	**Technical Challenges and Reliability**: Implement failover mechanisms and redundancy.
Help and documentation	Provide user assistance	Offer accessible help and documentation that effectively guides users through anticipatory functionalities.	**User education and communication**: Maintain open channels for user queries and support.

Designers and researchers can use this table as a **diagnostic tool** during product evaluation. It helps frame usability testing, heuristic reviews, and internal audits around the distinct risks and responsibilities of AI-driven services.

17.3 CHECKLIST FOR DESIGN TEAMS: EVALUATING ANTICIPATORY EXPERIENCES

To make these heuristics actionable, we offer a **practical checklist**. This tool helps teams systematically assess whether anticipatory systems are transparent, fair, adaptable, and user-centered. Use the checklist:

- During **prototyping** to surface early design risks
- During **usability testing** to guide observation and discussion
- During **post-launch audits** to ensure that systems continue to align with user needs

TABLE 17.2   Proposed Anticipatory Design Heuristics Checklist.

Keep users informed Ensure transparency by keeping users aware of system operations, predictions, and underlying processes.	☑ **Provide real-time feedback**: Ensure the system clearly communicates its status and current operations (e.g., predictions in progress, actions being taken). ☑ **Educate users on data privacy**: Include visible and easy-to-access information about data privacy and system security practices. ☑ **Implement feedback loops**: Regularly update users on how their feedback has improved system performance.
Align with mental models Design anticipatory systems to align with users' expectations and familiar workflows.	☑ **Use intuitive language**: Ensure that the system uses terminology and interaction patterns familiar to the target audience. ☑ **Align with mental models**: Verify that the system's predictions and recommendations align with user expectations and behaviors. ☑ **Promote critical thinking**: Integrate design elements that encourage users to evaluate system predictions critically.
Allow user intervention Provide mechanisms for users to customize, adjust, or override system recommendations	☑ **Enable override options**: Provide users with clear ways to adjust or reject system recommendations. ☑ **Allow customization**: Include settings for users to tailor the system's behavior to their preferences (e.g., level of automation, notification preferences). ☑ **Provide feedback channels**: Allow users to provide input on the system's predictions and actions.

TABLE 17.2 (Continued)

Maintain design coherence Ensure consistency in design elements and interaction patterns across all anticipatory features.	☑ **Maintain design coherence**: Ensure consistent design elements across all anticipatory features (e.g., colors, fonts, button placements). ☑ **Validate models regularly**: Update models to reflect current and accurate data, ensuring that predictions remain relevant.
Minimize anticipatory errors Incorporate safeguards to reduce the likelihood of erroneous predictions or recommendations.	☑ **Minimize erroneous predictions**: Use diverse and high-quality datasets to train algorithms. ☑ **Test system robustness**: Conduct thorough testing to identify potential biases or inaccuracies in predictions. ☑ **Provide preemptive warnings**: Notify users about potential inaccuracies or limitations in predictions.
Reduce cognitive load Present anticipatory information in ways that minimize the mental effort required from users.	☑ **Contextualize recommendations**: Present recommendations in a way that connects to the user's current context, reducing cognitive load. ☑ **Explain predictions clearly**: Include explanations of why specific predictions or actions are being suggested. ☑ **Highlight system benefits**: Provide accessible, clear information about how the system benefits users.
Cater to user proficiency Design systems to accommodate both novice and expert users.	☑ **Accommodate diverse users**: Ensure that the system is usable by both novice and expert users with different levels of technical proficiency. ☑ **Balance automation and control**: Integrate human oversight into critical automated processes. ☑ **Streamline workflows**: Design anticipatory processes that enhance user efficiency without introducing unnecessary complexity.
Present information cleanly Ensure that the anticipatory interface is aesthetically pleasing and focused on delivering essential insights.	☑ **Focus on essential information**: Present recommendations and system feedback in a clean and uncluttered format. ☑ **Prioritize clarity**: Avoid overwhelming users with excessive or irrelevant details. ☑ **Use visual hierarchy**: Ensure that important elements stand out visually while maintaining a cohesive design.
Assist users in error resolution Provide mechanisms to help users understand and recover from errors in anticipatory interactions.	☑ **Provide recovery options**: Offer users clear pathways to resolve errors caused by the anticipatory system. ☑ **Implement failover mechanisms**: Ensure system reliability by adding redundancy to critical processes. ☑ **Guide users during errors**: Include step-by-step instructions for resolving common errors.
Provide user assistance Offer accessible help and documentation to guide users through anticipatory functionalities effectively.	☑ **Make help accessible**: Include a dedicated help section with detailed documentation on anticipatory system features. ☑ **Provide contextual assistance**: Offer help or tips relevant to the user's current interaction within the system. ☑ **Maintain support channels**: Ensure that users can easily reach out to customer support or submit feedback.

Each checklist item corresponds to a heuristic principle. Together, they form a lightweight yet robust evaluation framework. The goal is not to create a rigid scorecard but to spark critical conversations and guide thoughtful design decisions.

17.4 CLOSING REFLECTIONS: DESIGNING SYSTEMS THAT ANTICIPATE WITH CARE

Anticipatory systems are no longer speculative—they shape how we shop, work, learn, and care for ourselves. With that influence comes a higher standard of responsibility. These heuristics are not static rules; they are evolving guardrails for a rapidly changing terrain.

As new capabilities emerge—from multimodal prediction to fully autonomous interaction—our evaluation methods must evolve too. But at their core, great anticipatory systems share a simple goal: **to be helpful without being harmful, intelligent without being intrusive**. By grounding anticipatory design in rigorous heuristics, we can move beyond hype and build systems that not only anticipate our needs but earn our trust.

17.5 DESIGNING WHAT COMES NEXT

Anticipatory systems don't just predict what we'll do next—they shape our behavior. As designers, this places us not in a position of control, but of deep responsibility and accountability. Every proactive suggestion, every invisible nudge, becomes part of the ongoing dialogue between human intent and machine interpretation.

This book has offered tools, frameworks, and heuristics—but the real challenge lies ahead, in the fluid and unpredictable context of real people's lives. Designers, researchers, and technologists must continue to ask: what futures are we reinforcing with each AI-powered interaction? What assumptions are we embedding into the systems we create? And how do we design technologies that respond without overreaching?

These aren't rhetorical questions. They're design prompts for an evolving field—questions that demand rigor, humility, and ongoing critical reflection. As we build systems that anticipate, we must also make space for ambiguity, autonomy, and disagreement. The goal isn't to predict perfectly—it's to design responsibly, knowing that the future we shape belongs to the people who live in it. As I noted in the heuristics, **leave room for the unexpected**. Great systems don't just anticipate the user—they accommodate them.

As you move forward, may these tools help you design systems that act not only with intelligence but with care. Anticipatory design isn't a trend. It's a responsibility.

TL;DR

Anticipatory systems require a new set of heuristics. This chapter adapts traditional usability principles to meet the unique demands of predictive, autonomous, and data-driven systems.

- Keep users informed: Make system behavior transparent and understandable.
- Align with mental models: Match system logic with user expectations.
- Allow user intervention: Provide clear options to override or adjust predictions.
- Minimize anticipatory errors: Design for reliability and error resilience.
- Reduce cognitive load: Present recommendations in context, not complexity.
- Balance control and automation: Support diverse user needs without overreach.

These heuristics form the foundation of a practical checklist that teams can use throughout the design lifecycle. The goal isn't perfection—it's principled progress toward systems that anticipate with care, clarity, and ethical responsibility.

References

1. H. Helskyaho, J. Yu, and K. Yu, *Machine Learning for Oracle Database Professionals: Deploying Model-Driven Applications and Automation Pipelines*. Springer, 2021, doi: 10.1007/978-1-4842-7032-5.
2. J. Krohn, G. Beyleveld, and A. Bassens, *Deep Learning Illustrated: A Visual, Interactive Guide to Artificial Intelligence*, 1st ed. Addison-Wesley Professional, 2019.
3. P. Dominé, *Music as Experience*. The Startup, Medium. Accessed: Apr. 14, 2025. Available: https://medium.com/swlh/music-as-experience-e1c18030c40a.
4. Available: https://commons.wikimedia.org/wiki/File:Detected-with-YOLO-Schreibtisch-mit-Objekten.jpg.
5. Available: https://blogs.microsoft.com/accessibility/seeing-ai-app-launches-on-android-including-new-and-updated-features-and-new-languages/.
6. Gartner, "What Generative AI Means for Business," Gartner, last modified [latest available date, e.g., Jun. 2, 2024, or n.d.]. Available: https://www.gartner.com/en/insights/generative-ai-for-business.
7. YouTube Saleforce 2020 Keynote. [Online]. Available: https://youtu.be/msNCdAjCaYs?si=YDJ9OOlkPmu4tZmS&t=4164.
8. H. Yuan and A. A. Hernandez, "User Cold Start Problem in Recommendation Systems: A Systematic Review," *IEEE Access*, vol. 11, 2023, pp. 136958–136977, doi: 10.1109/ACCESS.2023.3338705.
9. M. O. Riedl, "Human-Centered Artificial Intelligence and Machine Learning," *Hum Behav Emerg Technol*, vol. 1, no. 1, 2019, pp. 1–8. [Online]. Available: http://arxiv.org/abs/1901.11184.
10. Available: https://venturebeat.com/entrepreneur/zebra-medical-vision-raises-30-million-to-help-radiologists-detect-diseases-with-ai/.
11. Available: https://apps.apple.com/in/app/wysa-mental-health-ai/id1166585565.
12. Available: www.tesla.com/autopilot.
13. E. Pariser, *The Filter Bubble: What the Internet is Hiding from You*. Penguin Press, 2011.
14. D. Vacek, "Two Remarks on the New AI Control Problem," *AI and Ethics*, Sep. 2023, doi: 10.1007/s43681-023-00339-9.
15. Adapted from Stanford d.School Designing Machine Learning Course. [Online]. Available: https://designwith.ml/.
16. Available: https://play.google.com/store/apps/details?id=com.google.android.apps.seekh.
17. YouTube Google's Centered Ep. 9. [Online]. Available: https://youtu.be/fIlbN6ZixnE?si=Tsy5u8RaVHx3vnKG&t=361.
18. C. Saunders, *Emotional and Cognitive Overload*. New York: Routledge, 2019.
19. B. Schwartz, *The Paradox of Choice, Why More is Less*, 2004. [Online]. Available: http://wp.vcu.edu/univ200choice/wp-content/uploads/sites/5337/2015/01/The-Paradox-of-Choice-Barry-Schwartz.pdf.
20. Available: www.indiewire.com/2016/07/netflix-decide-watch-studies-1201708634/.
21. S. Tran, *Netflix Design Patterns and Flows*. Medium, UX Collective, Jan. 21, 2021. Available: https://sarahvitran.medium.com/netflix-patterns-and-flows-a5f1f76e3e09.
22. S. Fisher, "How to Rent Movies on Amazon," *Lifewire, Tech for Humans*, last modified Dec. 8, 2023. Available: https://www.lifewire.com/how-to-rent-movies-on-amazon-5192135.

23. D. Hirshleifer, Y. Levi, B. Lourie, and S. H. Teoh, "Decision Fatigue and Heuristic Analyst Forecasts," *Journal of Financial Economics*, vol. 133, no. 1, 2019, pp. 83–98, doi: 10.1016/j. jfineco.2019.01.005.

24. M. Twenge, *iGen: Why Today's Super-Connected Kids Are Growing Up Less Rebellious, More Tolerant, Less Happy–and Completely Unprepared for Adulthood.* Atria Books, 2017.

25. J. Kelly, "Indeed Study Shows That Worker Burnout Is At Frighteningly High Levels: Here Is What You Need To Do Now." Accessed: Aug. 18, 2022. [Online]. Available: https:// www.forbes.com/sites/jackkelly/2021/04/05/indeed-study-shows-that-worker-burnout-is-at-frighteningly-high-levels-here-is-what-you-need-to-do-now.

26. A.-L. Le Cunff, *Tiny Experiments: How to Live Freely in a Goal-Obsessed World*, 1 ed. New York: Penguin Random House, 2025.

27. A. Shapiro, "The Next Big Thing In Design? Less Choice." Accessed: Apr. 22, 2018. [Online]. Available: https://www.fastcompany.com/3045039/the-next-big-thing-in-design-fewer-choices.

28. I. P. Levin, S. L. Schneider, and G. J. Gaeth, "All Frames Are Not Created Equal: A Typology and Critical Analysis of Framing Effects," *Organ Behav Hum Decis Process*, vol. 76, no. 2, 1998, pp. 149–188, doi: 10.1006/obhd.1998.2804.

29. Available: https://www.scribbr.com/research-bias/framing-effect/.

30. R. Poli, "The Many Aspects of Anticipation," *Foresight*, vol. 12, no. 3, 2010, pp. 7–17, doi: 10.1108/14636681011049839.

31. G. Pezzulo and M. V. Butz, *The Challenge of Anticipation: A Unifying Framework for the Analysis and Design of Artificial Cognitive Systems.* Springer, 2008, doi: 10.1108/k.2010.06739eae.001.

32. N. Peckham and B. Berkebile, "Comprehensive Anticipatory Design Science: The Vision of R. Buckminster Fuller," in *Second Annual GreenBuild International Conference & Expo*, Pittsburgh, PA, 2003, pp. 1–6. [Online]. Available: www.bfi.org.

33. F. Galdon and A. Hall, "Prospective Design: A Future-Led Mixed-Methodology to Mitigate Unintended Consequences," *The International Association of Societies of Design Research (IASDR)*, 2019.

34. Available: https://www.webdesignmuseum.org/gallery/netflix-in-2006.

35. Available: www.statnews.com/2018/07/25/ibm-watson-recommended-unsafe-incorrect-treatments/.

36. W. Xu, M. J. Dainoff, L. Ge, and Z. Gao, "From Human-Computer Interaction to Human-AI Interaction: New Challenges and Opportunities for Enabling Human-Centered AI," in *2021 IEEE International Conference on Systems, Man, and Cybernetics (SMC)*. IEEE, 2021, pp. 1262–1267.

37. A. Kore, *Applied UX Design for Artificial Intelligence Designing Human-Centric AI Experiences.* Apress, 2022. [Online]. Available: https://link.springer.com/.

38. S. Raisch and S. Krakowski, "Artificial Intelligence and Management: The Automation-Augmentation Paradox," *Acad Manage Rev*, 2020, pp. 1–48, doi: 10.5465/2018.0072.

39. M. Kaushik, M.-T. Huang, A. Turner, S. Varanasi, and V. Grajski, *UXAI: A Visual Introduction to Explainable AI for Designers.* UXAI, 2020. Accessed: July 29, 2025. Available: https://www.uxai.design/ai-basics.

40. J. Cerejo, "Redefining UX: Behavior and Anticipatory Design in the Age of AI," *UXPA Magazine*, 2025. https://uxpamagazine.org/redefining-ux-behavior-and-anticipatory-design-in-the-age-of-ai/.

41. Available: https://info.vanta.com/hubfs/23-24%20Checklists%20-%20New%20Brand/ ISO42001_Compliance_Checklist.pdf.

42. Available: www.oecd.org/en/topics/ai-principles.html.

43. B. J. Fogg, "A Behavior Model for Persuasive Design," in *Proceedings of the 4th International Conference on Persuasive Technology*, 2009. Accessed: Dec. 18, 2024, doi: 10.1145/1541948.1541999.

44. Available: www.mealprep.com.au/p/noom-for-weight-loss-a-nutritionists-verdict.
45. R. Thaler and C. Sunstein, *Nudge: Improving Decisions About Health, Wealth, and Happiness*. Yale University Press, 2008.
46. P. Lally, C. H. M. van Jaarsveld, H. W. W. Potts, and J. Wardle, "How are Habits Formed: Modelling Habit Formation in the Real World," *European Journal Social Psychology*, vol. 40, no. 6, 2010, pp. 998–1009, doi: 10.1002/ejsp.674.
47. M. Celi and C. Colombi, "Trends as Future Prompts in the Anticipatory Design Practice," *Futures*, vol. 121, Aug. 2020, doi: 10.1016/j.futures.2020.102564.
48. G. Cascini, "TRIZ-Based Anticipatory Design of Future Products and Processes," *J Integr Des Process Sci*, vol. 16, no. 3, 2012, pp. 29–63, doi: 10.3233/jid-2012-0005.
49. A. Morrison, P. Dudani, B. Kerspern, and A. Steggell, "Amphibious Scales and Anticipatory Design," Aug. 2021, doi: 10.21606/nordes.2021.18.
50. S. Malhotra, L. K. Das, and V. M. Chariar, "Design Research Methods for Future Mapping," in *International Conferences on Educational Technologies 2014 and Sustainability, Technology and Education*, 2014, pp. 121–130. [Online]. Available: http://files.eric.ed.gov/fulltext/ED557342.pdf.
51. C. Kaya and A. Öner, *Anticipation: Meaning and Usage*. Yeditepe Üniversitesi Yayınevi, 2019.
52. A. M. Walorska, "Turning Data Into Experiences. Pro-active Experiences and their Significance for Customers and Business," *Procedia Manuf*, vol. 3, no. Ahfe, 2015, pp. 3406–3411, doi: 10.1016/j.promfg.2015.07.531.
53. J. Ask, S. Powers, R. Koplowitz, R. Warner, A. Stewart, and R. Birrell, "Anticipatory Experiences: The Challenges Anticipating Customers' Needs and Serving Them Proactively in Their Moments is Extremely Difficult," in *Report no. RES163315*. Cambridge, MA: Forrester Research, Inc., 2025. Available: https://www.forrester.com/report/anticipatory-experiences-the-challenges/RES163315-.
54. T. Zamenopoulos and K. Alexiou, "Towards an Anticipatory View of Design," *Des Stud*, vol. 28, no. 4, 2007, pp. 411–436, doi: 10.1016/j.destud.2007.04.001.
55. T. K. Gandhi et al., "How Can Artificial Intelligence Decrease Cognitive and Work Burden for Front Line Practitioners?," *JAMIA Open*, vol. 6, no. 3, Oct. 2023, doi: 10.1093/jamiaopen/ooad079.
56. P. Humes, "Cognitive Overload in the Contact Center is Costing You More Than You Think." ICMI, May 20, 2024, last updated Sept. 26, 2024. Available: https://www.icmi.com/resources/2024/ccaas-cognitive-overload.
57. S. Makridakis, S. Wheelwright, and R. Hyndman, *Forecasting: Methods and Applications*, 3rd ed. Hoboken, NJ: Wiley, 2008.
58. J. Pearl and D. Mackenzie, *The Book of Why: The New Science of Cause and Effect*. Basic Books, 2018.
59. Available:https://www.kickstarter.com/projects/hdrop/hdrop-real-time-hydration-wearable-device-monitor.
60. S. Kleber, "How to Get Anticipatory Design Right." [Online]. Available: www.hugeinc.com/articles/how-to-get-anticipatory-design-right.
61. Available: https://newsroom.spotify.com/2019-06-12/your-daily-drive-music-and-news-thatll-brighten-your-commute/.
62. Available: https://www.youtube.com/watch?v=3NWMWFHOrRA&ab_channel=FFChannel.
63. R. Miller and D. Gugerli, *Transforming the Future: Anticipation in the 21st Century*, 1st ed. Routledge, 2018.
64. R. Poli, *Introduction to Anticipation Studies*. Trento, Italy: Spring International Publishing, 2017. [Online]. Available: www.springer.com/series/15713.
65. R. Miller, "Anticipation: The Discipline of Uncertainty," *Futures of Futures*, 2012, pp. 39–43. [Online]. Available: https://drive.google.com/open?id=0B7Bn-eBPZZX7Ty1fS3JsMTFqbjQ%0Awww.apf.org/page/Publications.
66. D. Norman, *The Design of Everyday Things*. New York, USA: Basic Books, 1986.

67. T. Zamenopoulos and K. Alexiou, "Collective Design Anticipation," *Futures*, vol. 120, Nov. 2018/2020, p. 102563, doi: 10.1016/j.futures.2020.102563.

68. R. Rosen, *Anticipatory Systems: Philosophical, Mathematical, and Methodological Foundations*, 2nd ed. New York, USA; Oxford: Pergamon Press, 1985/2012, doi: 10.1016/j.prro.2011.06.014.

69. J. Lovejoy, "The UX of AI, Using Google Clips to Understand How Human-Centered Design Process Elevates Artificial Intelligence," *Google Design*, Jan. 25, 2018. Available: https://design.google/library/ux-ai.

70. P. Daugherty and J. Wilson, *Human + Machine: Reimagining Work in the Age of AI*. United States: Harvard Business Review Press, 2018.

71. M. Geden et al., "Construction and Validation of an Anticipatory Thinking Assessment," *Front Psychol*, vol. 10, Dec. 2019, pp. 1–10, doi: 10.3389/fpsyg.2019.02749.

72. B. Rollier and J. A. Turner, "Planning Forward by Looking Backward: Retrospective Thinking in Strategic Decision Making," *Decision Sciences*, vol. 25, no. 2, 1994, pp. 169–188, doi: 10.1111/j.1540-5915.1994.tb01838.x.

73. Available: www.sweetpotatotec.com/salesforce-einstein-ai-vs-human-expertise-striking-the-balance.

74. H. J. Einhorn and R. M. Hogart, "Decision-Making: Going Forward in Reverse," *Harv Bus Rev*, vol. 20, no. 4, 1987, p. 125, doi: 10.1016/0024-6301(87)90169-5.

75. GCPSE–Global Centre for Public Service Excellence, Foresight Manual-Empowered Futures for the 2030 Agenda. Singapore: United Nations Development Programme, 2018.

76. J. Voros, "Big History and Anticipation," in *Handbook of Anticipation*, P. Roberto, Ed. Cham: Springer International Publishing, 2017, pp. 425–464, doi:10.1007/978-3-319-31737-3_95-1.

77. G. Fasoli and R. Poli, "Future Orienteering Evaluation Model: Forecasting, Foresight and Anticipation Indices," *European Journal of Futures Research*, vol. 12, no. 1, Sep. 2024, p. 19, doi: 10.1186/s40309-024-00242-4.

78. R. Poli, "Foresight," in *The Palgrave Encyclopedia of the Possible*, V. P. Glăveanu, Ed. Cham: Springer International Publishing, 2022, pp. 577–583, doi: 10.1007/978-3-030-90913-0_76.

79. R. Poli, *Handbook of Anticipation, Theoretical and Applied Aspects of the Use of Future in Decision Making*, 2019, doi: 10.1007/978-3-319-91554-8_65.

80. J. Cerejo and M. Carvalhais, "Anticipation as a Tool for Designing the Future," in *Advances in Design and Digital Communication IV*, N. Martins and D. Brandão, Eds. Cham: Springer Nature Switzerland, 2024, pp. 37–52.

81. G. A. Miller, "The Magical Number Seven, Plus or Minus Two: Some Limits on Our Capacity for Processing Information[1]," *Psychol Rev*, no. 63, 1956, pp. 81–97. [Online]. Available: http://psychclassics.yorku.ca/Miller/.

82. "How to Do Horizon Scanning: A Step-by-Step Guide," *Futures Platform*, Oct. 2021. Available: https://www.futuresplatform.com/blog/how-to-horizon-scanning-guideline.

83. J. Nielsen, "Usability Heuristics," in *Usability Engineering*, M. Kaufmann, Ed. Cambridge: Academic Press Inc., 1994, ch. 5, pp. 115–163.

Index

For Product Safety Concerns and Information please contact our EU
representative GPSR@taylorandfrancis.com
Taylor & Francis Verlag GmbH, Kaufingerstraße 24, 80331 München, Germany